PREVENTING FAMINE

PREVENTING FAMINE

Policies and prospects
for Africa

*Donald Curtis, Michael Hubbard,
and Andrew Shepherd*

*With contributions from Edward Clay, Hugh
and Catherine Goyder, Barbara Harriss,
Richard Morgan, and Camilla Toulmin*

Routledge
London and New York

First published in 1988 by
Routledge
11 New Fetter Lane, London EC4P 4EE

Published in the USA by
Routledge
in association with Routledge, Chapman & Hall, Inc.
29 West 35th Street, New York NY 10001

Printed in Great Britain by
Biddles Ltd, Guildford and King's Lynn

British Library Cataloguing in Publication Data
Curtis, Donald
   Preventing famine: policies and prospects
   for Africa.
   1. Famines——Africa——Prevention
   I. Title II. Hubbard, Michael
   III. Shepherd, Andrew
   363.8   HC800.Z9F3
   ISBN 0–415–00711–9
   ISBN 0–415–00712–7 Pbk

Library of Congress Cataloging in Publication Data
Curtis, Donald, 1939–
   Preventing famine.
   Includes bibliographies and index.
   1. Famines——Africa——Case studies. 2. Food relief——Africa
   ——Case studies. 3. Agriculture and state——Africa. 4. Food
   supply——Africa. I. Hubbard, Michael. II. Shepherd, Andrew.
   III. Title.
   HC800.Z9F326 1988   363.8'7'096   87–28267
   ISBN 0–415–00711–9
   ISBN 0–415–00712–7 (pbk.)

Contents

Notes on the Contributors

Edward Clay is a Fellow of the Institute of Development Studies at the University of Sussex and Director of the Relief and Development Institute in London. He has worked as an economist at the Bangladesh Rice Research Institute and has been an adviser to the Bangladesh Agricultural Research Council. His previous academic experience was as a Lecturer in Economics at the University of Papua New Guinea. He is on the editorial board of *Food Policy* and is convenor of the Study Group on Food Aid of the UK Development Studies Association. Recently he has been researching the food aid policies of the EEC, the United States, and recipient countries in Africa.

Donald Curtis is a Senior Lecturer in the Development Administration Group, Institute of Local Government Studies, University of Birmingham. He has a particular interest in local organization and development. His overseas experience is extensive and includes work in the Sudan, Botswana, and Gujarat State, India.

Catherine Goyder has a special interest in urban community development and has been involved in urban community development projects in Hyderabad, India, and in Addis Ababa, Ethiopia, where she lived from 1982–6.

Hugh Goyder worked for OXFAM as their representative in Central India from 1979–82 and was Country Representative for OXFAM in Ethiopia from 1982–6. He now works in the Research and Evaluation Unit of OXFAM in Oxford.

Barbara Harriss has recently joined Queen Elizabeth House, Oxford, as a lecturer in agricultural economics, having had her interest in famine prevention and poverty alleviation policies kindled by six years in the Department of Human Nutrition of the London School of Hygiene and Tropical Medicine. With field experience dating from 1969 in South Asia and later in Sahelian West Africa, her special interests are in the grain trade, intrahousehold resource allocations and welfare, and the implementation of food policies.

Michael Hubbard is a Lecturer in the Development Administration Group, Institute of Local Government Studies, University of Birmingham. His main interest is in agriculture and international trade. Much of his research experience is in Botswana where he taught economics at the University of Botswana.

Richard Morgan worked as National Food Strategy Co-ordinator in Botswana from 1983–5, and subsequently as a Consultant in the SADCC Food Security Programme. Since July 1986 he has been head of UNICEF's Emergency Programme in Mozambique. His contribution, written in 1986, was made in a private capacity.

Andrew Shepherd is a Lecturer in the Development Administration Group, Institute of Local Government Studies, University of Birmingham. He is at present on secondment to UNICEF as a field officer in the Sudan, to which country he has made numerous visits as a lecturer and researcher since 1978.

Camilla Toulmin is a research officer at the Overseas Development Institute, working on problems of arid land development and irrigation management. She is particularly interested in the environmental aspects of livestock and crop production in the Sahelian region of Africa.

Preface

'Why can't something be done about it?' was the question from a colleague back in 1985 at the height of the news coverage of the latest famine in Africa. He was probably just expressing the frustration – and supposing that we shared the frustration – of being a spectator at the end of a one-way channel of communication when something is clearly going wrong that one knows could be different.

We responded in part because, having been involved for some years in teaching, studying, and sometimes taking more active roles in rural development, we each had *some* ideas about how drought can be handled and famine avoided. We knew that within the profession there were more. But what struck us very forcibly was the failure of this available knowledge to show itself in the debate of the early 1980s about economic policies and development strategies for Africa.

After the Sahelian drought in the early 1970s a plethora of reports and recommendations had emerged, several research programmes had been undertaken and projects started. However, in the policy-generating centres for Africa the resulting material must have been filed away under some such heading as 'Desertification' while more pressing problems such as debt and balance of payments remained on the desk. For instance, there was nothing in the recommendations of the Berg Report (World Bank 1981), supposedly the basis of a new consensus, that indicated that resilience against famine was a proper objective of economic policy. So to do something towards putting famine consciousness back upon the economic policy agenda was for us a second motive. The recurrence of famine in parts of Ethiopia and Sudan in late 1987 re-emphasizes the urgent need for effective counter-measures.

A workshop which we held in Birmingham revealed that these concerns were widely shared amongst fellow professionals in the fields of rural development and agriculture. There was no uniformity of views amongst the twenty or more participants and some substantial disagreements, particularly perhaps about the relationship between the kinds of policies which were accepted as necessary and broader aspects of economic policy. However, there was much common ground and several participants contributed directly to a paper entitled 'Africa post famine' that was presented in the summer of 1985 to the British minister of overseas development, several international agencies, non-government organizations, and to the annual conference of the Development Studies Association.

Encouraged by this workshop and again drawing widely upon direct contributions from experienced academics and practitioners, we launched upon this book. While the aim of the workshop paper was to make a point about what we saw as bias in policy, this book is intended to summarize experience in dealing successfully with the threat of famine as well as describing shortcomings that lead to failure. It illustrates the short-term measures that can be applied to counter famine threats as well as some of the long-term preventative policies. The scope of the chapters is inevitably wide-ranging. Because of the complexities of the social and ecological environments which face the risk of famine prescriptions cannot always be offered. The message is nevertheless simple. If the migrations and starvation of the early 1980s which followed the similar events of the early 1970s are not to be repeated in the early 1990s, a consciousness of famine risks should be allowed to shape all aspects of public policy – food policy and aid between countries as well as agriculture and rural development within countries.

We would like to think that we have maintained some of the ideas of the 1985 Birmingham workshop in this text and hereby acknowledge the stimulus which the participants provided to this work.

Our direct contributors are acknowledged in the contents list and in the text. They were all asked to prepare material on the basis of their experience and we have incorporated their work substantially as we received it. We have, however, exercised some editorial discretion in order to maintain a consistency of theme and presentation. This has hopefully not led to any errors of fact or interpretation.

For the non-attributed material we share the responsibilities of authorship. We have all worked over each of these chapters and have agreed to live with the results as they now stand. Nevertheless, any reader who wishes to discuss specific points may be helped by knowing that the first drafts of Chapters 1 and 2 as well as the Gujarat case study (in Ch. 5) were prepared by Michael Hubbard. The Sudan study (Ch. 3) and the technology chapter (Ch. 8) were drafted by Andrew Shepherd, while Chapters 9 and 10 fell to Donald Curtis to draft.

Royalties earned on all copies sold will be donated to Oxfam as a contribution to famine prevention.

PART I

BACKGROUND AND CASE STUDIES

1
Introduction

'The era in which we live has the odious distinction of being the period when more people will die of famine than in any previous century' (Cahill 1982:1). Yet in no previous era has famine been more preventable than the present; through unexcelled communication and transport possibilities and world grain reserves, the international capability exists to detect and remedy food shortages anywhere before famine results. Social, not natural or technological, obstacles stand in the way of modern famine prevention.

This marks modern famines as man-made to a greater extent than those of previous times. Greater population pressure on the land, with frequently adverse ecological consequences reducing long-term productivity, fewer natural 'fall back' possibilities (e.g. migrating to better land), greater dependence of local production and consumption on goods sold to and bought from afar, have all reduced the ability of poor people to withstand adversity by using their own resources. They now rely more heavily on governments (often remote from them) for relief. War as a man-made cause of famine has intensified as the scale of wars has expanded while their frequency remains undiminished.

Each major famine is followed by a flood of academic literature analysing why it happened, how it can be avoided in future, and (in some cases) apportioning blame. The 1973–4 famine in Ethiopia and the Sahel produced acute international concern over desertification in the region. But a dozen years after this and after the creation of the United Nations Environment Programme, there is greater famine-proneness in the region and a less favourable aid and trade climate. The urgent need for rethinking must not be daunted by the knowledge that this has been done many times before.

This book stems from such rethinking by a group of mainly British-based people working in universities and voluntary organizations who met at Birmingham University in 1985. The meeting took place against the background of the world-wide publicity of the famine and relief effort in Africa and the much quieter drama being enacted in the offices of donor organizations, where a new and tougher consensus – led by the World Bank's Berg Report (World Bank 1981) – on Africa's worsening development and debt problems had emerged. Debt relief and development funds were being made increasingly conditional upon cutting government budget deficits, devaluation, and privatization.

We would like to believe that much which emerged from the workshop and has been developed further into this book is of practical value in famine prevention. All the contributors are people who implement policy in Africa and elsewhere, mainly on rural development and food security. As such, their emphasis is on what is feasible and effective in famine prevention, not on trying to set up an alternative theoretical consensus. In some areas, particularly with regard to long-term measures for famine prevention, there are no clear models and much need for an experimental and innovative approach.

There are two major concerns which emerge clearly. First, current attention to Africa's economic problems focuses strongly on debt and increasing exports in order to overcome balance-of-payments deficits. The preoccupation with current finance, important as it is, has hijacked much of the attention and funds previously devoted to rural development. But it is rural development, with its stress on asset creation and protection for the poor, though it produces low financial returns in the short term, that holds the key to effective famine prevention in the long term. This is particularly the case for people living in arid and semi-arid areas, who are the ones most at risk of famine. This concern has to do with what can be called 'livelihoods' – not simply incomes – implying some command over productive assets as well as food and cash, and with greater 'security against impoverishment'.

Second, as regards emergency action to prevent famine, there is much that can be learnt from the administrative methods used by countries (such as India, Bangladesh, and Botswana) which have become relatively successful in recent years in famine prevention, though not necessarily in reducing the underlying vulnerability of the population. That the Indian model is less salutary and less applicable to African conditions as regards long-term famine prevention measures is a theme of Chapter 6.

The famine process

It is useful to identify three categories of cause of famines. First, there are *long-term* causes of household income loss or income instability which increase the vulnerability of poor people. High among these in Africa is environmental degradation, affecting pastoralists in particular but also cultivators in arid and semi-arid areas. The Sahel, the Horn and western southern Africa are the areas where long-term factors have been increasing famine risk most. Social changes, particularly increased assetlessness among rural people, also increase famine risk. Among these are occupation of the best land by the rich and consequent loss of access by the poor and the breakdown of traditional social obligations to the poor.

Failure to prevent famine, but only to relieve it after it is substantially advanced, is probably the greatest single cause of increased assetlessness among rural people in drought-prone environments. Major droughts are times of great 'shaking out' in the rural social structure. Livestock, jewellery, even land, are sold at ruinous prices in the desperation to survive; fortunes can be made in a few months by the buyers. To be saved by famine relief is to come away with virtually nothing but your life. Long-term measures for famine prevention are therefore those which increase and protect the assets of the poor.

Second, there are *precipitating* factors, the events which dislodge the last food security of the poor, setting off the secondary events which worsen the situation – spiralling food prices, collapsing prices of rural assets (particularly livestock, because of lack of feed and the need to sell), calling in of debts, laying-off of employees, ultimately abandonment of the aged, the sick and the very young, and migration to towns. Precipitating factors include all those which actually reduce the food supply (drought, floods, war, epidemics) as well as those which it is feared will do so, and any which reduce the purchasing power of the poor. The great famine in Bengal in 1943 was created by rapid inflation, vigorous speculation, and panic hoarding in anticipation of further price increases in a time of war, plus administrative chaos through failed procurement, abolition of wholesale price control, and restriction of grain movement which prevented inflows into Bengal. Sen argues that the result was severe loss of 'exchange entitlement' (purchasing power) by the poor, despite no overall shortage of food in the country (Sen 1981).

Third, there is *relief failure.* For famine to be precipitated, governmental famine-prevention administration must be inadequate, incompetent, or unable to operate. In Bengal 'administrative chaos' (ibid.: 76) in the famine relief operation worsened the famine. In Bangladesh, improvement in famine relief administration during the 1970s greatly reduced famine in the early 1980s (see Ch. 5). Effective relief has prevented famine in Botswana and Gujarat state in India (see Ch. 5). Delays in relief have been a major factor in recent famines in Sudan and Ethiopia (see Ch. 3). The politics of famine relief (international and national) have played their part in worsening the Ethiopian and Mozambique famines through delaying the delivery of food aid (Tickner 1985: 91) while civil wars have severely hampered distribution.

Famine is therefore a process beginning with the existence of a large number of people living so close to subsistence that any disruption that reduces their purchasing power or their access to resources from which they derive their livelihood threatens many of them with starvation. 'The basic failure in the understanding of famine we have today', concludes Rangasami, 'is the inability to recognise the political, social and economic determinants that mark the onset of the process' (Rangasami 1985). This lopsided condition in the administration of famine prevention emerges graphically in this book: much experience has been accumulated and lessons learnt in the administration of emergency relief in different countries; much less has been achieved and learnt in the administration of long-term measures to remove underlying causes of famine.

Famine-prevention policy

Much stress has been laid recently on the importance of preventing collapse of real income, following Sen (1981) and the Indian Famine Codes. This emphasis has been developed further by the World Bank in its publication *Poverty and Hunger*, which attempts to map the policy options for governments of developing countries trying to increase food security for their populations (World Bank 1986a). It stresses the correlation between low income levels of countries and the extent of malnutrition in those countries. Comparing national figures, 'Income growth was the largest single influence on dietary improvement between 1970 and 1980' (ibid.: 3).

It argues that the main policy decisions on long-term food security are, first, whether to produce food or import it; second, whether to subsidize food prices to consumers; and third, to what extent to augment incomes directly.

Targeting becomes an issue in administering food price subsidies and income augmentation schemes. In general, the more that subsidies or income generation are targeted to the poor (in feeding schemes, 'fair price' shops, asset distribution and employment creation), the more administratively intensive they are; the less they are targeted (especially if made market-wide), the more financially costly they become. Thus it is argued that there is a trade-off between financial and administrative costs in the decision to target or not.

The World Bank's prescription for reducing long-term food insecurity in countries regarded as typical of Sub-Saharan Africa (the poorest people living in the rural areas, still largely dependent on agriculture and livestock for their incomes) is the strengthening of agricultural production, by raising producer prices of internationally traded products to border price levels, and improving the technical, input, and marketing support to agriculture, for the production of both traded and non-traded agricultural commodities. Additionally, subsidizing the retail price of processed commodities where they are widely consumed by the poor (e.g. flour) may, it argues, be cost-effective (automatic targeting).

By contrast, where most of the poor are urban (therefore not food producers), held to be typical of several South American countries, the policy mix should include mainly consumer subsidies targeted to the poor, administered so that producer prices are not reduced. Where the poor are widely distributed through the urban and rural areas (i.e. both as net producers and consumers of food, held to be typical of Asia) a combination of these two approaches is recommended by the Bank.

That increasing food production must be a major part of policy to increase food security in many African countries is borne out in Chapter 2, which details the trend towards decreasing food production per head in several countries, combined with shortages of foreign exchange, and the bleak prospects for increasing foreign exchange earnings in the medium term. Thus the import option as a means of increasing aggregate food availability is largely closed (other than where food aid is readily available in the quantities and qualities required).

But there is also in part a trade-off between increasing non-food agricultural exports (e.g. cotton) and increasing national food production, since they compete for the same land, especially the high-grade land. For example, farmers in the state-controlled irrigation schemes in Sudan had a long struggle to be allowed to respond to the demand for *dura* (sorghum), expressed in high market prices for this staple food, while the authorities tried to maintain the output of cotton and other historically important exports.

The stress laid by the Berg Report on increasing cash crop exports (World Bank 1981) has now been dropped from World Bank official statements. But balance-of-payments problems and pressure on governments from creditors negotiating debt-restructuring conditions, to maximize export earnings (despite declining export markets, it seems) reduces their scope for switching their research, marketing and land resources towards food production for domestic consumption.

Further, a long-term policy of increasing food production and subsidizing key commodities consumed by the poor is insufficient to reduce the vulnerability of the people in Africa whose food security is most at risk, that is pastoralists and small farmers and their dependants in remote and arid areas. It is not clear that their food security has been increased by many large-scale crop production schemes, particularly where they have lost access to land and gained little employment. They have little or no political power and, as a result, benefit little from government expenditures. The arid, semi-arid, or mountainous environments where they live are wasting rapidly through deforestation and erosion. Their crop and livestock production has benefited little from government research and extension.

We feel these issues must be at the top of the agenda in any discussion of famine prevention. They are the focus of Chapters 6 to 10 in which the policy options arising from the case studies of famine and famine prevention in Part I are discussed.

Structure of the book

Our approach has been to start by trying to identify which countries in Africa are most at risk of famine and why (Ch. 2). We then look in detail at cases of failure and success in preventing famine. The studies of two neighbouring but very different countries, the Sudan (Ch. 3) and Ethiopia (Ch. 4), provide an historical perspective on the circumstances surrounding famine as well as an analysis of events leading to the recent occurrences. The studies show that it is often at times of political instability that difficult climatic conditions combine with inadequate governmental responses to trigger famine. The accounts of recent events bear out the same picture.

The accounts in Chapter 5 from Botswana and the state of Gujarat in India – which read, comparatively speaking, as success stories – show, in some detail, how administrative systems can be built up to deal with emergencies just as severe as those in Sudan and Ethiopia. These got far less press coverage because human migrations and starvation were not allowed to happen. In the same chapter, from a slightly different perspective, the Bangladesh account shows the links between this kind of emergency provision and other aspects of state action to achieve food security. The news here is good. It should be read carefully to see how relevant these emergency systems are for other countries which have not coped as well.

In Chapter 6 the relevance of Indian subcontinental experience is directly addressed, and it makes the point that this experience is not only in handling emergencies. Equally important have been the whole range of measures which have led, on the one hand, to increased investment in agriculture and growing output and, on the other, to food security: getting food to the needy by administrative means. Both contributors to this chapter stress that the systems devised are far from ideal but nevertheless work in the Indian subcontinental context. Both raise questions about the relevance of such systems to Africa.

Two points of convergence emerge from the case study based material. The first is the necessity for some administrative capability if emergency situations are to be managed. The second is that there needs to be a depth of famine-relevant policy in agriculture, livestock, and infrastructure development, as well as food security provisions, if risks are to be minimized.

Policy development in livestock and in agricultural technology is taken up in Chapters 7 and 8. The development possibilities and famine-resistance requirements of the vast but very varied drought-prone areas of Africa are very different from those of the flat, irrigable or well-watered areas of the world where the 'green revolution' and other now well-tried technologies have made their impact. Differences in population density, social structure, and control of productive resources also present challenges. In technology development, and also in relation to the control of productive resources (addressed in Chapter 9), we find that it is not possible to generalize about what constitutes an appropriate technology or a successful means of securing for 'at risk' categories of farmers an adequate title to land, or water, or livestock. However, in both cases we can point to some useful lines of development and to a process through which policy can support the identification of good technologies or social-entitlement practices.

This leaves the final chapter (Ch. 10) to tackle the question of how relief activities, food security measures, and a range of policies and programmes that counter famine risk are to be carried out. Our approach is not to say simply that this or that must be done, but to look for indications of strengths within government agencies, political institutions, non-government organizations, and, not least, local communities, that can be drawn upon to achieve particular ends. The emphasis is upon making the most of what is available now rather than awaiting some better-funded, better-trained, politically harmonious future.

This last remark points to an issue which recurs throughout the book, particularly in considering the relevance of the 'Indian model' to Africa. Can one anticipate more external aid funds for African countries to counter the effects of debt burdens and deteriorating terms of trade, or do African countries face the prospect of further financial stringencies? The extent of need is addressed in our Chapter 2, but it is not possible to predict whether expected shortfalls will be met by aid. The availability of funds in government coffers will constrain both the range of programmes which can be fielded to counter famine and also the administrative capacity to deal with them. If we could anticipate continual growth in public-sector budgets and in trained staff then the 'Indian model', which is administratively intensive, might be expected to transport more readily to African environments. If decline is more realistic, the hunt should be on for less administratively intensive measures. In the latter chapters of the book we have tended towards the view that policy development should not assume that expanding resources will be the norm. But all is not gloom. Several of the policy options available, such as helping to re-establish social controls over grazing, or promoting 'sideways extension' among farmers, may actually be cheaper than alternatives. Much can be done by economizing. There are virtues in being obliged to raise local resources. It would nevertheless be better still if these options could be taken up *and* the additional funds made available to develop transport and communication – the infrastructure necessary for food security.

2
Famine and the National and International Economy

This chapter attempts to throw some light on the following questions:

1. To what extent have world recession, falling commodity prices, severe debt and national macroeconomic policies been responsible for increasing poverty and famine in Africa?
2. Will adequate finance be available for famine prevention in Africa in coming years?
3. Which African countries are going to be in most need of external assistance for famine prevention and relief in future?

The role of macroeconomic factors in famine risk and famine prevention

Changes in the national economy (public spending, inflation, import capacity, employment growth) probably play an even more important role in modern famine generation and prevention than they did in past eras. Even where macroeconomic factors have not precipitated famine they can worsen it through inflation, poor alternative employment opportunities for those threatened with starvation and inadequate administrative, infrastructural and financial means for detecting and relieving hunger.

In general, macroeconomic factors affect *long-term famine risk* via the availability or non-availability of private or public finance for investment in infrastructure, services, production, reclamation and reafforestation in the disadvantaged areas, which are usually the ones with the highest famine risk. Such investment is long term, promises no quick returns and is therefore not commercially attractive; it also suffers heavily in times of macroeconomic stress.

Economic development has been acutely uneven in much of Africa, with investment concentrated in cities and those rural areas most favoured by existing roads and railways. Policies which have raised urban incomes primarily (urban bias in public expenditure, subsidies on imported foods, heavy taxation of marketed agricultural products, monopolistic marketing systems which pay only a minor fraction of earnings to the farmers, overvalued exchange rates which cheapen imports and reduce the value of exports) have been both a cause and a result of low investment in rural areas, particularly the less favoured areas. Long-term declines in the international prices of some agricultural commodity exports (like tea, sisal, and cotton) reduce incomes and employment in the 'cash crop' zones, reduce tax revenue, and reduce the seasonal employment on which some of the rural poor from non-cash crop zones heavily depend (e.g. in Sudan, see Ch. 3)

While macroeconomic factors can *precipitate* famine (through rapid inflation or large-scale loss of employment), they are probably more likely (particularly in the relatively remote places in which recent African famines have taken place) to be critical in determining whether a precipitating factor (such as drought) is allowed to set off the chain of income and asset losses which make a famine. Import and transport capacity (to prevent spiralling food prices), alternative employment opportunities and lines of credit (to prevent income collapse), adequately funded health facilities to prevent the spread of disease which is the killer among the debilitated – all these are crucial. Most importantly, the *funding of effective famine prevention and relief* is determined largely by state revenues and therefore ultimately by the macroeconomic health of the economy.

In sum, the relation between famine and the national economy is obviously many-sided and is intermediated by several other factors – politics, geography, administration – but two general points can be made:

1. *Famine can and does occur in the midst of plenty.* Therefore it is not to be expected that famine will take place only when aggregate food supplies in relation to population fall, since famine is localized in particular groups and areas which have suffered a major loss of income. This assumption is now well substantiated (Sen 1981, World Bank 1986a, Van Apeldoorn 1980) on the mid-1970s drought in Nigeria and the Sahel, and many commentaries on famines in India, e.g. *The Gujarat Relief Manual*, Vol. 1 (Government of Gujarat Revenue Department 1976).

2. *Where there are large famine-vulnerable groups in the country, substantial losses of national income and of import capacity increase famine risk by weakening both long-term famine prevention and the relief effort.* Specifically:

(i) Employment and purchasing power of pay are reduced.

(ii) Import capacity is likely to be reduced (therefore there is less ability to import food for relief).

(iii) State budgets per capita tend to be smaller (therefore poorer infrastructure of roads and services and greater likelihood of inadequate famine relief effort and of inadequate attention to arresting soil erosion and deforestation).

(iv) More rural people tend to be driven by hardship into seeking a living at the expense of natural resources (e.g. forests, easily eroded hill slopes) thereby further undermining future production.

Point (2) is the one to be argued here. It amounts to claiming that aggregate falls in purchasing power (including import capacity) increase famine risk where they reduce either long-term famine-prevention efforts or famine relief. The approach taken is to identify African countries regarded as having high famine risk and to look at their recent record on food consumption and production, income and import capacity growth, to assess whether these factors have increased famine risk or not, and whether they are likely to do so in the future.

Which are the famine-vulnerable countries in Africa and why?

Twenty-seven African countries have been variously identified by relief agencies – UN Office for Emergency Operations in Africa, the Food and Agriculture Organization (FAO), and World Food Programme (WFP) – as critically affected by recent droughts and facing severe food shortages by the mid-1980s (Table 2.1). They include countries recently prominent in news headlines above pictures of famine-ravaged rural people (Ethiopia, Sudan, Mozambique), others scarcely mentioned in the world's media (São Tomé and Príncipe, Cape Verde). This list of countries containing people at risk from famine (henceforth the 'at risk' countries) is certainly not exhaustive but provides a broad base for analysis.

Tables 2.1 to 2.8 assemble data, mainly of the twenty-seven 'at risk' countries, on trends in national food consumption, food production, source of food, income levels and growth and import capacity. Although the figures are all drawn from official sources – as published by FAO, International Monetary Fund (IMF), and World Bank – many are probably little more than estimates. For what they are worth, they suggest the following:

Food consumption

Consumption of calories per capita per day in all these 'at risk' countries together has been on an upward trend from 1964/6 to 1980/2 (Table 2.1). But individually some of them have experienced sizeable declines: Burkina Faso (7 per cent), Burundi (6 per cent), Ghana (17 per cent), Kenya (9 per cent), Mali (16 per cent), Mozambique (7 per cent). Burkina Faso, Ghana, Mali and Mozambique had all fallen below 2,000 calories by 1980–2. Figures for Angola, Chad and Guinea are not available.

Average calorie consumption per day for the African continent as a whole in 1980–2 (2,391) was above that for the twenty-seven countries critically affected by drought (2,190). The overall trend in Africa as a whole, and in other continents, is to increases in per capita calorie intake. But the increases in Africa, however, are less than in Asia and Latin America, and on present trends, by the early 1990s Africa will have replaced Asia as the continent with the lowest per capita calorie intake (Table 2.2).

Food production

The decline in per capita food production in Africa has been widely publicized in the 1980s. Among the 'at risk' African countries this decline has been at an average compound rate of 1.8 per cent since 1970. This is the result of population growth of some 2.8 per cent, outrunning food production growth of 1.3 per cent.

Although the increases in Africa's food production from the mid-1970s to early 1980s were substantially less than those of Asia and South America, they were greater than the average of all developed countries. But with Africa experiencing the highest population growth in the world, a further decrease of 10 per cent in per capita food production is in prospect by 1994, whereas all other regions will have increased per capita production, particularly Asia (Table 2.4).

Table 2.1 Food supply: calories per capita per day in 'at risk' countries (all sources)

	1964–6	1974–6	1980–2
Angola	n.a.	n.a.	n.a.
Botswana	1,963	2,079	2,468
Burkina Faso	2,074	1,973	1,922
Burundi	2,393	2,266	2,244
Cape Verde	1,795	2,216	2,716
Central African Republic	2,053	2,241	2,151
Chad	n.a.	n.a.	n.a.
Ethiopia	n.a.	n.a.	n.a.
Gambia	2,345	2,152	2,223
Ghana	2,004	2,155	1,657
Guinea	n.a.	n.a.	n.a.
Guinea Bissau	1,979	2,173	2,230
Kenya	2,241	2,177	2,036
Lesotho	2,041	2,057	2,355
Mali	2,089	1,926	1,749
Mauritania	2,048	1,802	2,186
Mozambique	2,007	1,978	1,864
Niger	2,131	1,984	2,462
Rwanda	1,705	1,923	2,115
São Tomé and Príncipe	2,139	1,872	2,351
Senegal	2,397	2,226	2,364
Somalia	2,096	2,067	2,077
Sudan	1,848	2,150	2,332
Tanzania	1,982	2,255	2,409
Togo	2,242	2,059	2,160
Zambia	2,092	2,320	2,124
Zimbabwe	2,095	2,132	2,164
ALL	2,076	2,095	2,190

Source: *FAO Production Yearbook*, 1984, Table 105.
Note: n.a. = figure not available.

Table 2.2 Food supply by continent: calories per capita per day (all sources)

	1980–2	1990–2*
Africa	2,391	2,496
North and Central America	3,319	3,436
South America	2,609	2,692
Asia	2,379	2,560
All developing	2,388	2,554
All developed	3,395	3,545
World	2,652	2,782

Source: *FAO Production Yearbook*, 1984, Table 105.
*Projections from 1964–6, 1969–71, 1974–6, 1980–2.

Table 2.3 Food production in 'at risk' countries in Africa (1969–71 = 100)

	1974–6		1982–4	
	Production	Per capita	Production	Per capita
Angola	100	89	103	72
Botswana	124	110	106	71
Burkina Faso	106	95	122	92
Burundi	112	100	121	92
Cape Verde	n.a.	n.a.	n.a.	n.a.
Central African Republic	115	104	128	97
Chad	90	83	103	80
Ethiopia	103	91	125	93
Gambia	113	103	101	78
Ghana	103	90	90	61
Guinea	94	86	102	78
Guinea Bissau	100	100	117	86
Kenya	103	87	120	74
Lesotho	107	98	95	71
Mali	88	78	114	83
Mauritania	76	69	91	66
Mozambique	100	89	99	65
Niger	89	77	126	89
Rwanda	115	90	166	99
São Tomé and Príncipe	n.a.	n.a.	n.a.	n.a.
Senegal	128	114	111	76
Somalia	100	88	113	60
Sudan	126	108	135	92
Tanzania	111	92	141	89
Togo	70	61	79	57
Zambia	125	107	118	78
Zimbabwe	126	107	113	73
ALL	105	92	114	79

Source: FAO Production Yearbook, 1977, Tables 4 and 6; 1984, Tables 4 and 9.
Note: n.a. = figure not available.

Table 2.4 Regional food production projections to 1994 (1974–6 = 100)

	1982–4		1994	
	Production	Per capita	Production	Per capita
Africa	115	90	139	81
North and Central America	117	104	146	115
South America	128	107	169	120
Asia	130	112	168	126
All developing	130	110	171	123
All developed	112	105	128	112

Source: FAO Production Yearbook, 1984, Tables 4 and 9. Own projection.

The adverse trend in per capita food production has, in the case of most of the 'at risk' countries since the mid-1970s, been accompanied by massive increases in cereals imports (35 per cent of total cereals consumed in 1984), much of it in the form of food aid (35 per cent of cereals imported in 1982) (Tables 2.5 and 2.6) and the virtual disappearance of cereals exports (except from Burkina Faso, Sudan and Zimbabwe).

Table 2.5 Volume of cereals imports and exports of 'at risk' countries: annual average during 1982–4 (1975–7 = 100)

	Exports	Net imports	Net imports as % of total consumed (1984)*
Angola	0	250	52
Botswana	0	196	85
Burkina Faso	129	252	8
Burundi	0	192	4
Cape Verde	0	180	96
Central African Republic	0	302	21
Chad	0	379	14
Ethiopia	0	673	10
Gambia	0	128	47
Ghana	0	169	26
Guinea	0	237	26
Guinea Bissau	0	128	25
Kenya	55	†	24
Lesotho	0	222	54
Mali	0	305	27
Mauritania	0	194	89
Mozambique	0	171	40
Niger	0	1023	2
Rwanda	0	162	6
São Tomé and Príncipe	0	168	91
Senegal	0	171	50
Somalia	0	286	48
Sudan	282	405	20
Tanzania	0	132	13
Togo	0	419	23
Zambia	0	176	21
Zimbabwe	800	− 691	16

Sources: FAO Production Yearbook, 1977, Tables 9 and 15; *FAO Trade Yearbook*, 1984, Tables 35 and 36.
*Net grain imports / (grain production + net grain imports).
†Kenya was a net exporter in the mid-1970s but became a substantial net importer in the early 1980s.

Table 2.6   Cereals food aid as a percentage of cereals imports and production in 'at risk' countries

	Cereals imports		Cereals production	
	1975	1982	1975	1982
Angola	0	23	0	20
Botswana	11	11	9	37
Burkina Faso	0	n.a.	0	7
Burundi	66	44	2	2
Cape Verde	17	n.a.	138	n.a.
Central African Republic	6	7	1	2
Chad	126	43	2	5
Ethiopia	5,590	64	1	3
Gambia	58	59	10	19
Ghana	51	22	6	8
Guinea	73	43	7	8
Guinea Bissau	50	87	12	18
Kenya	3	42	0	4
Lesotho	33	31	7	28
Mali	95	47	11	5
Mauritania	49	49	126	320
Mozambique	18	41	6	19
Niger	397	56	9	4
Rwanda	112	90	9	4
São Tomé and Príncipe	0	n.a.	0	n.a.
Senegal	13	16	4	10
Somalia	95	44	53	43
Sudan	40	47	2	7
Tanzania	32	78	10	9
Togo	0	8	0	2
Zambia	1	49	1	13
Zimbabwe	n.a.	0	n.a.	0

Sources: FAO Production Yearbook, 1977, Tables 9 and 15; *FAO Trade Yearbook*, 1984, Tables 35 and 36; World Bank, 1981, Table 24; 1981, Table 20.
Notes: Data are not strictly comparable with respect to time or source: food aid is for the growing season, cereals imports are for the calendar year; import figures are from importing countries, food aid figures are from donors.
n.a. = figure not available.

Falling food production per capita and rising imports of grains to maintain aggregate consumption has been a common pattern among all the 'at risk' countries for which figures are available, during the 1970s to early 1980s (Table 2.7). But not all have maintained consumption levels: notably Burkina Faso, Burundi, Ghana, Kenya, Mozambique – all of which, except Mozambique, have a below-average ratio of cereals imports to total cereals production and have suffered falling import capacity, except Burundi (see Tables 2.9 and 2.11). Therefore falling import capacity may possibly be partly responsible for their falling consumption levels; but this must remain speculation given the undoubtedly wide margin of error in the data.

This shift in the source of food has probably had a distributional impact. The steady to rising aggregate calorie consumption levels

Table 2.7 Food imports as a percentage of total merchandise imports

	*1965**	*1978**	*1982**
Angola	18	n.a.	n.a.
Botswana	n.a.	n.a.	n.a.
Burkina Faso	25	19	25
Burundi	18	23	n.a.
Cape Verde	n.a.	n.a.	n.a.
Central African Republic	13	17	n.a.
Chad	13	15	n.a.
Ethiopia	7	6	10
Gambia	24	24	n.a.
Ghana	13	9	n.a.
Guinea	n.a.	n.a.	n.a.
Guinea Bissau	n.a.	43	n.a.
Kenya	n.a.	7	8
Lesotho	n.a.	n.a.	n.a.
Mali	21	19	n.a.
Mauritania	9	n.a.	n.a.
Mozambique	17	n.a.	n.a.
Niger	13	10	24
Rwanda	12	n.a.	n.a.
São Tomé and Príncipe	n.a.	n.a.	n.a.
Senegal	37	23	27
Somalia	33	25	20
Sudan	24	19	19
Tanzania	n.a.	11	7
Togo	18	16	26
Zambia	10	6	9
Zimbabwe	7	2	n.a.
ALL	17	16	18

Sources: World Bank 1981, Table 9; 1986b, Table 9.
Notes: *Many of the figures are for the closest available year.
          n.a. = figure not available.

(whether they are accurate or not) probably mask shifts in consumption patterns, the assumption being (and there are not the data to support it) that incomes, and therefore consumption, have generally held up better in the towns than the countryside.

The dramatic rises in cereals imports of the early to mid-1980s are partly the result of the great trans-continental drought which devastated production. But the pre-existing common trend to increasing reliance on food imports is the result of disruptions to the economy (war, droughts in the 1970s), rapid population growth, falling productivity through expansion of cultivation into more arid areas, plus failure of agricultural policies and macroeconomic policies which subsidized food imports and made commercial food production less attractive.

Incomes and import capacity

The shift in the source of food from home to foreign sources has been accompanied by falls in income and in international purchasing power among many of the 'at risk' countries, particularly in the early 1980s.

On average, real GNP per capita fell during 1982–4 for the 'at risk' countries in aggregate, almost to its level of 1979 (Table 2.8), which for some (Niger, Ghana, Senegal, Zambia) was the continuation of a downward trend established over the previous decade (World Bank 1986a: 67). These countries experiencing long-term falls in GNP per capita also suffered long-term contraction of public and private spending and investment during 1973–83 (Table 2.8), as did Burkina Faso and Central African Republic.

As regards import capacity (Table 2.9) the data available indicate that all 'at risk' countries (except Botswana, Central African Republic, Chad, Lesotho) suffered losses on their trade in goods and services during 1978–84, which were not made up by inflows of long-term capital (i.e. current account and long-term capital account negative). This indicates that they were relying on short-term capital (import credits probably, as well as rolled over or defaulted-upon debt service obligations) in order to sustain their imports or prevent them falling further.

Of the 'at risk' countries for which data are available several did manage to maintain or increase their imports per capita during 1973–83 (Angola, Botswana, Central African Republic, Ethiopia, Gambia, Mali, Niger, Rwanda, Togo) but among these some were doing so at the expense of their future consumption: Gambia, Mali, Niger, Togo are

Table 2.8 GNP per capita, public and private spending and gross investment in 'at risk' countries

	GNP per capita (1979 = 100)		Public consumption spending (% annual growth)	Private consumption spending (% annual growth)	Gross domestic investment (% annual growth)
	1982	1984	1973–83	1973–83	1973–83
Angola	n.a.	n.a.	n.a.	n.a.	n.a.
Botswana	125	126	13.7	7.8	5.7
Burkina Faso	116	89	3.6	1.7	−3.7
Burundi	156	122	5.4	2.8	15.7
Cape Verde	n.a.	n.a.	n.a.	n.a.	n.a.
Central African Republic	107	93	−1.5	3.2	−6.7
Chad	73	n.a.	n.a.	n.a.	n.a.
Ethiopia	108	85	7.1	2.6	7.1
Gambia	144	104	n.a.	n.a.	n.a.
Ghana	151	146	4.8	−1.3	−8.1
Guinea	111	107	6.4	2.0	−0.7
Guinea Bissau	100	106	2.2	1.7	3.5
Kenya	103	79	6.3	3.6	3.4
Lesotho	150	156	n.a.	n.a.	n.a.
Mali	129	100	7.5	2.8	4.2
Mauritania	147	141	1.4	3.0	7.0
Mozambique	n.a.	n.a.	n.a.	n.a.	n.a.
Niger	115	70	2.3	6.6	3.5
Rwanda	130	135	n.a.	n.a.	n.a.
São Tomé and Príncipe	n.a.	n.a.	n.a.	n.a.	n.a.
Senegal	114	88	6.6	3.3	−0.7
Somalia	n.a.	90*	1.5	7.9	−8.2
Sudan	119	92	4.5	7.6	5.6
Tanzania	108	81	n.a.	3.0	4.4
Togo	97	71	8.4	3.3	−0.2
Zambia	128	94	−0.8	3.9	−0.5
Zimbabwe	181	157	10.8	2.9	1.9
ALL	125	107	5.0	3.6	1.7

Sources: World Bank 1981, 1984, 1986b.
Notes: *1982 = 100; no figure for 1979.
n.a. = figure not available.

among the twelve African countries suffering the most severe long-term debt problems (along with Benin, Liberia, Madagascar, Somalia, Sudan, Tanzania, Zambia) according to the World Bank (1986b: 53). Doubt hangs over their ability to sustain import capacity in the future.

Table 2.9   Import capacity of 'at risk' countries

	Annual growth of $ value of merchandise trade 1973–83 (%)		Annual growth of terms of trade 1970–84 (%)	Current account + long-term capital (annual average) (SDR millions)	
	Exports	Imports		1978–81	1981–4
Angola	8.3	3.3	−4.7	n.a.	n.a.
Botswana	n.a.	n.a.	n.a.	27	39
Burkina Faso	1.7	4.2	−1.8	−6	−18
Burundi	n.a.	n.a.	n.a.	n.a.	n.a.
Cape Verde	n.a.	n.a.	n.a.	n.a.	n.a.
Central African Republic	3.8	2.5	1.7	4	17
Chad	−3.1	−8.6	1.3	0	17
Ethiopia	1.4	2.7	−4.7	−52	−29
Gambia	−6.9	5.0	−1.5	n.a.	n.a.
Ghana	−6.4	−8.0	−1.1	−31	−61
Guinea	n.a.	n.a.	n.a.	n.a.	n.a.
Guinea Bissau	n.a.	n.a.	n.a.	n.a.	−10
Kenya	−4.8	−4.6	−0.9	−260	−128
Lesotho	n.a.	n.a.	n.a.	10	13
Mali	5.1	3.9	−1.0	−31	−81
Mauritania	0.5	−0.8	−4.5	−9	−47
Mozambique	−8.3	−4.2	−1.9	n.a.	n.a.
Niger	19.0	11.5	−5.2	−103	n.a.
Rwanda	2.6	12.9	−0.4	2	−17
São Tomé and Príncipe	n.a.	n.a.	n.a.	n.a.	n.a.
Senegal	−0.9	−1.2	−1.3	−97	n.a.
Somalia	7.3	0	−2.6	−40	−43
Sudan	−1.5	1.3	−0.9	−222	−189
Tanzania	−4.6	−2.7	−1.6	−201	n.a.
Togo	3.5	7.4	n.a.	−7	−9
Zambia	−0.8	−7.3	−8.7	−174	−203
Zimbabwe	n.a.	n.a.	n.a.	−128	−245
ALL	0.8	0.9	−2.0	−69	−58

Sources: World Bank 1986, Tables 7 and 11; IMF 1985, Table A3.

The question which arises is whether falling import capacity is linked substantially to famine risk or not. In short, would it be possible, were it the priority of government, to shift import content into goods for famine relief (food, medicines, fuel, personnel) and long-term famine prevention (much less import intensive than famine relief), even in the face of falling import capacity?

Dealing with famine relief first, it is clear that self-financing of the large emergency flows of food necessary could involve large increases in import costs. First, food aid is already a large proportion of cereals imports (see Table 2.6) in many of the 'at risk' countries. Second, that proportion increases many fold in times of drought (e.g. mid-1970s). Taking Ethiopia as an example, imports of cereals and preparations amounted in 1982 to approximately 7 per cent of merchandise imports. Adding cereals food aid would bring this proportion up to about 11 per cent. Together cereals imports and cereals food aid amounted to about 8 per cent of cereals production in 1982. But with the onset of drought and famine, cereals food aid increased hugely, to about one-third of production in 1984. Self-financing of this would have increased the total bill for merchandise imports substantially.

The impact of limited import capacity on long-term causes of famine reveals itself in restricted development budgets, shortages of vehicles and fuel, deteriorating roads and communications, eroding real value of civil servants' pay and morale – all of which restrict the ability to mount development programmes or to respond quickly in an emergency in remote areas. A comparison of Botswana (see Ch. 5) and Sudan (see Ch. 4) illustrates the difference which political stability, strong export revenues, and relatively good management make to effective famine prevention.

Ranking of the 'at risk' countries according to famine risk

Tables 2.10 and 2.11 rank the 'at risk' countries roughly and somewhat subjectively in descending order of famine risk, with war taken as the factor causing highest famine risk, followed by high percentage of population vulnerable to drought, low GNP per capita level and whether GNP per capita fell 1979–84, and stagnant or falling import capacity.

What emerges from this rough classification is that countries which are most vulnerable to drought are not necessarily the ones with highest famine risk, although the Sahelian countries are generally high risk. Chad, Ethiopia and Sudan are vulnerable on all the indicators and therefore rank as the most at risk. Burundi, Zimbabwe, and Lesotho rank the least at risk.

Table 2.10 Indicators for ranking of famine-vulnerable countries

	1 War	2 % population, drought vulnerability high	3 Low-income country	4 1984 GNP per head <1979	5 Import capacity stagnant or falling
Angola	*			?	?
Botswana		*			
Burkina Faso		*	*	*	*
Burundi			*		
Cape Verde		*	?	?	?
Central African Republic		*	*		
Chad	*	*	*	*	*
Ethiopia	*	*	*	*	*
Gambia		*	*	?	?
Ghana			*		*
Guinea		*?	*		
Guinea Bissau		*?	*		*
Kenya			*	*	*
Lesotho					
Mali		*	*		*
Mauritania		*	*		*
Mozambique	*		*	?	*
Niger		*	*	*	*
Rwanda			*	?	
São Tomé and Príncipe		?	?	?	?
Senegal		*	*	*	*
Somalia		*	*	*	*
Sudan	*	*	*	*	*
Tanzania			*	*	*
Togo			*	*	*
Zambia				*	*
Zimbabwe					*

Notes: * = indicator applies to this country.
space = indicator does not apply to this country.
? = information not available.

Columns:
1 *Assuming war currently increases famine risk in the country.*
2 *Rough estimate based on proportion of country which is arid and proportion of population which is rural.*
3 *According to classification in World Bank 1986.*
4 *From Table 2.8 above.*
5 *Import capacity classified as decreasing or stagnant if terms of trade 1973–83 falling, or current account + long-term capital 1981–4 negative, or negative rate of growth of imports and exports 1973–83 (see Table 2.8).*

Financing famine prevention in Africa

With famine caused by many different factors, finance alone is not enough. But without finance for both long-term and short-term measures, very little can be done. Four critical points regarding finance for famine prevention must be made.

First, the costs of famine prevention are substantial. In the recent (1982–4) drought in Botswana, the public works programme for drought relief alone cost 5 per cent of all government spending from own resources (see page 117). The state of Rajasthan in India in 1986 spent most of its development budget on preventing famine following one of the worst droughts experienced there.

Second, the costs of famine prevention are far lower than the costs of famine relief – even leaving aside the intolerable human and social cost of allowing famine to begin instead of preventing it. Preventing prices of food from rocketing, livestock prices from collapsing, supporting in good time the incomes of the able-bodied and of those not able to work – these involve less financial, administrative, and import cost than relief with its emergency feeding programmes, food distribution to households, vast medical cost in preventing epidemics and saving the lives of starving children, organizing camps for migrants who have abandoned their homes, rehabilitation and restocking. It is probably fair to say that the costs of full relief and rehabilitation are so great as to make its achievement unlikely in the extreme. Full restocking alone, at the hugely increased prices of livestock prevailing after the drought, is prohibitively expensive. As for other forms of wealth lost by the poor – for example, jewellery, – plans for rehabilitation are not even contemplated. Famines, if relieved rather than prevented, are times of great 'shaking out' in the social structure, with many of the poor becoming assetless and some of the rich increasing their fortunes quickly.

Third, in the countries with highest famine risk finance is least available for preventive measures. Projections of the financial prospects of most of the 'at risk' countries in the medium term are extremely bleak. The World Bank (1986b: Appendix A) has projected exports, net capital inflows, and likely debt repayment (assuming further rescheduling) for African countries and concludes that in twelve (including Sudan, Mali and Somalia) import capacity will 'decline sharply' during 1986–90 without additional debt relief or capital inflows. Other African countries eligible for assistance through the International Development Agency

(IDA) or the World Bank, (i.e. including all the 'at risk' countries except Botswana and Zimbabwe) are projected to experience an 11 per cent decline in per capita import capacity during 1986–90.

Summing up the picture as a whole for the IDA-eligible African countries, the World Bank concludes that with 'the maximum feasible effort on export earnings' external payments will amount to $35.3 billion per year during 1986–90 and inflows will be $32.8 billion per year, leaving an annual gap of $2.5 billion. These projections are based on external payments and inflows during 1980–2. 'Bridging the gap' of $2.5 billion, which is what World Bank argues should be the minimum target for donors, in addition to ensuring that they are not 'a net recipient of resource flows from an African country which is undertaking credible reform programs' (World Bank 1986b: 42) *would only restore import capacity to its level of 1980–2*, a level which for the most 'at risk' countries was wholly inadequate for famine prevention or relief. The conclusion to be drawn is that most of the 'at risk' countries will be heavily dependent on donors in the coming years for funds for famine prevention and/or relief.

Fourth, the organization of international funding for famine prevention and relief is unsatisfactory. Adequate finance for famine prevention has not been easily forthcoming, and funds and food for relief arrive too late – only when the pictures of starving children reached the western television screens did the millions start to flow in 1973–4 and in 1984–5. There is no reason to think that without a major change in the organization of funding for famine prevention the situation will be any different next time. What is required is commitment of funds and personnel to support the creation and maintenance of appropriate famine prevention programmes.

Table 2.11   Rough ranking of 'at risk' countries by present famine risk

---

1 *WAR DISRUPTED,*
plus:
High % of population vulnerable to drought, low income, poor import capacity:
  Chad
  Ethiopia
  Sudan
Low income, poor import capacity:
  Mozambique
Poor import capacity:
  Angola?

2 *HIGH PROPORTION OF POPULATION VULNERABLE TO DROUGHT*, plus:

Low income, poor import capacity:
Burkina Faso
Guinea Bissau
Mali
Niger
Senegal
Somalia
Cape Verde?
Gambia?
São Tomé and Príncipe?

Low income:
Central African Republic
Guinea

Lower middle income, good import capacity:
Botswana

3 *LOW INCOME, LOW IMPORT CAPACITY*
Ghana
Kenya
Rwanda
Tanzania
Togo
Zambia

4 *LOW INCOME:*
Burundi

5 *LOWER MIDDLE INCOME, POOR IMPORT CAPACITY:*
Zimbabwe

6 *LOWER MIDDLE INCOME*
Lesotho

Notes: The ranking is in terms of indicators in Table 2.9 above. The ranking is very rough: disruption by war is assumed to make for the highest risk, with proportion of population vulnerable to drought the next highest. Economic factors (income per capita, import capacity) are included as secondary.

3

Case Studies of Famine: Sudan
Andrew Shepherd

1984/5 will be recorded as the worst year in Sudan since 1888/9. The number of deaths due to starvation or malnutrition-related disease is not known. Save the Children Fund (UK) has put the number of deaths in Darfur, where food aid was least available, at between 50,000 and 150,000. (The population of Darfur is about 3.5 million.) No organization or individual has been willing to make an informed guess for the whole country. It must be a substantial number, particularly of children.

The overall verdict must be that famine was not averted, it was restrained; its impact was reduced by a massive foreign food aid operation. In particular, by 1986 this food aid distribution system was working sufficiently well to avert a repetition of 1985. 'Mass starvation has been averted' claimed the UN Emergency Operation in the Sudan (UNEOS) in October 1985.[1] In its review of the situation UNEOS overlooked the massive problems faced that year in Darfur, though it was more candid about the lesser but significant difficulties of Kordofan. In making its claim it also overlooked the experience of refugees from Ethiopia as well as Chad, perhaps because they were the responsibility of UNHCR.

The magnitude of the famine is difficult to assess. The only records of deaths came from the various camps. Thus in Maaskar el Ghaba (El Obeid) there had been nineteen deaths in a population of 22,000 between 8 November 1984 and 19 December 1984, that is roughly 0.01 per cent.[2] In the camps around Derudeib, in the Red Sea Hills, there were, according to UNICEF, between ten and twenty deaths per day in a population of 18,000, a much higher figure – 5.5 per cent if this situation were prolonged over 100 days.

In Darfur, de Waal estimated gross average annual mortality between

October 1984 and October 1985 at 7.6 per cent, varying from 3.1 per cent (slightly more than normal) in El Fasher town to 22.8 per cent in the far north, and from 3.3 per cent among nomads to 21.7 per cent among displaced people (de Waal *et al.* 1986: 71)

Table 3.1 Results of nutrition surveys of under-5-year-old children in Kordofan and Darfur, 1984 and 1985

Kordofan	September 1984			February 1985			May/June 1985		
	1	*2*	*3*	*4*	*5*	*6*	*7*	*8*	*9*
A	51	87.2	74.7	85.7	88.1	84.3	74.8	88.5	80.6
B	40	11.0	21.0	13.1	11.1	14.3	21.8	10.2	17.1
C	9	1.9	4.3	1.2	0.8	1.4	3.4	1.3	2.6

Darfur			May/June 1985		
	1	*2*	*3*	*4*	*5*
A	71	89	91	88	86.0
B	25	9	8	11	12.7
C	4	1	1	1	1.4

Sources: Oxfam, UNICEF and Kordofan Regional Government, *Nutrition-Surveillance and Drought Monitoring Project*, Reports of September 1984, February/March 1985 and May/June 1985. Oxfam, UNICEF and Darfur Regional Government: *Nutrition-Surveillance and Drought Monitoring Project: Report on Project Activities*, March–May 1985.

Key:
A   % with no or mild protein energy malnutrition.
B   % with moderate protein energy malnutrition.
C   % with severe protein energy malnutrition.

Kordofan:
1 Measured by mid-upper arm circumference (MUAC), Oxfam/ UNICEF.
2 Measured by weight for height (W/H), Ministry of Health, northern Kordofan only.
3 Measured by MUAC, Oxfam/UNICEF.
4 Measured by W/H, Oxfam/UNICEF.
5 Measured by W/H, Oxfam/UNICEF (final results).
6 Measured by W/H, Oxfam/UNICEF, northern Kordofan only.
7 Oxfam/UNICEF, northern Kordofan only.
8 Oxfam/UNICEF, southern Kordofan only.
9 Oxfam/UNICEF, all Kordofan.

Darfur:
1 Oxfam/UNICEF, northern Darfur sandy areas.
2 Oxfam/UNICEF, northern Darfur alluvial areas.
3 Oxfam/UNICEF, southern Darfur sedentary.
4 Oxfam/UNICEF, southern Darfur nomads.
5 Oxfam/UNICEF, all Darfur.

Nutrition surveys were carried out by Oxfam, UNICEF, and the Sudanese Ministry of Health in Kordofan and Darfur. These figures show how a very serious situation had developed in northern Kordofan and the sandy soil areas of northern Darfur by May/June 1985, and are reproduced in Table 3.1.

As far as Kordofan is concerned, the critical series of figures are those in columns 5 and 9, from which a marked deterioration in nutritional status between February and May/June can be seen. Columns 2, 6, and 7 give the comparison for northern Kordofan alone. Of children less than 5 years old, 25.2 per cent were moderately or severely malnourished in May/June compared with 15.7 per cent in February. What is remarkable from these figures is how little the situation changed between September and February (columns 2 and 5), the period before the major food aid distributions. The critical state of children in the sandy areas of northern Darfur also stands out.

Unfortunately, not very much can be learned from such data, although in practice, *faute de mieux*, they have to be the means of determining food needs and allocations. The main problem in this case is that we have little idea of what is normal, of what children would have weighed in a non-famine year at the same time. Second, the category of severely malnourished may be continually eroded if these children die in a famine; and normally the highest mortality rate would be expected in this category. Third, the categories A, B, and C are not universally accepted. Fourth, the picture may be distorted if large numbers of children are 'bunched' statistically near one of the dividing lines. Fifth, the two techniques used to measure nutritional status gave very different results (compare 1 and 2; 3 and 4). Mid-upper arm circumference measurements suggested that 49 per cent of the children were malnourished in September 1984, and served to focus attention on the situation. But this technique was later abandoned for practical purposes (Oxfam, February/March 1985b: 16). Despite all these difficulties, the figures did generally confirm commonsense observation; the worst affected were those unable to migrate. Within northern Kordofan, the east was the worst affected district in May/June 1985, despite its proximity to the source of food aid for western Sudan – Kosti. Some areas such as Um Gidad, north of Wad Banda, on the other hand, were distinctly better off, in this case due to irrigated vegetable production. Any area with a source of income less dependent on rainfall would have been better off, at least until grain prices shot up beyond most people's reach.

Generally people who migrated to Omdurman or the Nile and then returned or were persuaded to return to Kordofan were worst off of all. In Darfur the large numbers in informal camps around the towns tended to be worst affected nutritionally. What is interesting and as yet unresearched is how so many children (70–90 per cent) could survive so well in such a situation.

In the case of nomads it seems likely that those with large herds, from which they can consume and sell, would have staved off destitution longest. Some of them came through the drought and were able to benefit from high livestock and low grain prices in 1985 and 1986. In the case of sedentary cultivators, the few with large grain stores and livestock would have been best off. These would probably have been in good grain growing areas, and would have been families with favourable producer: consumer ratios, that is families with adult children.

In all cases, access to a non-agricultural source of income (e.g. weaving, wood-cutting, urban or government employment) and to wild foods would have been critical factors, especially in the absence of food aid or sufficient food aid. Several reports comment on the importance of wild foods. In northern Kordofan *mukhayt (Boscia Senegalensis)* fruit was a staple for many by February 1985. It was said to give children diarrhoea (Oxfam February/March 1985: 26) In southern Kordofan, with its lusher environment, famine foods were more easily available. In northern Darfur by June *mukhayt* was reported to have been exhausted (Oxfam June 1985: 8), but other famine foods were available (de Waal *et al.* 1986).

Cholera swept the country from east to west. Typhoid occurred in isolated outbreaks, for example at Um Badr, where there were twenty deaths from typhoid in the winter of 1984/5 (Oxfam February/March 1985:23).

Livestock died in large numbers. A report on the Eastern Region in September 1984 estimated losses there up to 40 per cent. A report on Darfur (Oxfam, June 1985) estimated that between 70 and 80 per cent of cows and goats had died in northern Darfur; between 40 and 50 per cent in southern Darfur (see Table 3.3) The figures for New Halfa and Darfur were based on questionnaires; others were guesstimates, and could be widely wrong, though intuitively this seems unlikely.

The process of distress sale of livestock had begun in 1984. The figures for the increases in sales of stock over that year in the country as a whole, according to the Livestock and Meat Marketing Corporation (LMMC), are given in Table 3.2. At this time the really dramatic

Table 3.2 Changes in livestock numbers sold through LMMC markets, December 1983 to December 1984

	Increase or decrease	*% change*
Cattle	3,197	2
Calves	16,296	54
Sheep	621,350	124
Goats	12,068	15
Camels	18,728	100
Horses	−1,383	−27
Donkeys	−6,474	−24

Source: LMMC Office, Kassala, October 1985.

increases in sales were in sheep and camels, animals of the semi-arid belt.

Livestock prices plummeted as a result. Table 3.3 shows the percentage change from 1984 prices, based on questionnaire data from Kordofan.

Table 3.3 Livestock prices in 1985 as a percentage of 1984 reported price, Kordofan region

	Population group							
	North sedentary		*South sedentary*		*Camel nomads*		*Cattle nomads*	
	Feb.	*May*	*Feb.*	*May*	*Feb.*	*May*	*Feb.*	*May*
Cattle	22.2	53.6	36.8	28.8	n.a.	n.a.	33.3	33.1
Sheep	24.1	46.8	41.4	19.1	40.6	20.5	20.5	23.3
Goats	29.1	71.3	41.4	32.4	35.3	26.9	35.7	31.3
Camels	12.5	48.7	n.a.	n.a.	20.5	42.2	n.a.	n.a.
Average	16.1	55.1	35.4	26.8	34.1	29.8	31.0	29.2

Source: Oxfam, May/June 1985: 16.
Note: n.a. = figure not available.

For sedentary people in northern Kordofan prices were reported as being one-sixth of 1984 prices in February; the proportion for all others was about one-third. By May 1985 prices in northern Kordofan had increased to half their 1984 value, reflecting the scarcity of livestock and possibly early rains bringing pasture growth. In later 1985 and 1986 livestock prices were two to three times their pre-drought level.

At the same time *dura* and *dukhn* prices soared out of the reach of ordinary people. In Kordofan (Table 3.4) and central Sudan (Ahmed 1985: 11 and 13) prices were two to three times their 1984 level and had more or less reached a high plateau by February 1985. In Darfur price levels per *malwa* (3.2 kg) were about two and a half times higher (Oxfam, June 1985: 32).

In the Northern Region *dura* was £S7.68/*malwa*, and wheat, the staple of the region, anything up to that amount (Ahmed 1985: 10). In Kassala Province prices increased fourfold (ibid.: 12).

Table 3.4 Reported prices of *dura* and *dukhn* 1984 and 1985, Kordofan region (£S/*malwa**)

| | Dura | | | Dukhn | | |
| | 1985 | | 1984 | 1985 | | 1984 |
	Feb.	May	(average)	Feb.	May	(average)
Northern Kordofan	4.39	5.03	1.76	5.97	6.48	2.38
Southern Kordofan	4.55	4.40	1.79	5.85	6.26	2.72
Kordofan as a whole	4.70	4.77	1.78	5.97	6.39	2.49

Source: Oxfam, May/June 1985: 17.
* 1 *malwa* = 3.2 kg.

These figures for livestock sales, prices and grain prices represent the classic famine squeeze in which the major asset that peasants, whether predominantly cultivators or pastoralists, can realize is livestock. This situation was very much to the benefit of the urban population in the short term, who could buy cheap meat with incomes which did not, by and large disappear, though for many they were reduced by the migration to the towns of numbers of work-seeking peasants. Incomes of the rural population are predominantly derived from crops and livestock. Widespread crop failure (Tables 3.5 and 3.6) annihilated or substantially reduced income from crops, whether grains or others, which would normally be sold. There was dramatically increased supply and thus reduced prices of firewood and charcoal as the major rural alternative income source. Drought typically accelerates deforestation. Wage-labouring opportunities were greatly reduced, especially in the peasant economy and rain-fed sector – the final straw which threw the peasants of western and eastern Sudan on the rough mercy of government and the international community.

Table 3.5 Crop production in the rain-fed sector 1984/5 (000s tonnes) indicating extent of crop failure in 1984/5

Province	Sorghum (1)	(2)	Millet (1)	(2)	Groundnuts (1)	(2)	Sesame (1)	(2)	Cotton (1)	(2)
Blue Nile	162	55	10	4	9	2	31	7	6	2
Gezira	38	4	1	1	n.a.	n.a.	n.a.	n.a.	n.a.	n.a.
White Nile	28	10	14	4	15	3	4	1	n.a.	n.a.
Northern Kordofan	58	8	103	20	161	11	52	6	n.a.	n.a.
Southern Kordofan	89	54	8	4	8	1	15	7	7	19
Northern Darfur	5	1	56	13	17	2	1	1	n.a.	n.a.
Southern Darfur	84	40	180	90	160	70	14	21	n.a.	n.a.
Southern Region	137	100	14	11	52	40	22	20	2	2

Source: Government of Sudan, Ministry of Agriculture and Natural Resources, *Crop Production 1985/6 Season: Preliminary Estimates*, Khartoum, November 1985.
Notes: These figures are at best good estimates. None is based on crop-cutting surveys.
Some (e.g. the increase of sesame output in Southern Darfur) are unlikely.
(1) = five-year average.
(2) = 1984–5.
n.a. = figure not available.

Table 3.6 1984/5 production as a percentage of five-year average

Province	Sorghum	Millet	Groundnuts	Sesame	Cotton
Blue Nile	40	40	22	23	33
Gezira	11	100	n.a.	n.a.	n.a.
White Nile	36	29	20	25	n.a.
Northern Kordofan	14	19	7	12	n.a.
Southern Kordofan	61	50	13	47	271
Northern Darfur	20	23	12	100	n.a.
Southern Darfur	48	50	44	150	n.a.
Southern Region	73	79	77	91	100

Source: See Table 3.5 above.
Note: n.a. = figure not available.

The social effects of famine were:

1. Vast migrations, temporary and permanent.
2. Further sedentarization of pastoralists.
3. Development of landlessness among families who migrated permanently from home.
4. Family breakdowns, leading to vagrancy.

These phenomena could be summarized as a dramatic acceleration in the stratification of peasant society. At the bottom a distinctive new class of assetless people, especially women and children, was created. Many peasant households were forced to rely much more on the labour market for their subsistence – a situation which was miserable in 1984/5 when there was little to harvest, but much better in 1985/6 when food aid reduced the supply of labour to the market dramatically and wages were higher. The latter, as much as food aid itself, must explain why malnutrition was significantly reduced except in the Red Sea Hills by October/November 1985 (Oxfam/UNICEF, October/November 1985). Many, if not most, rural households have part of their family in towns, in areas of agricultural surplus or abroad earning money. This money is not necessarily remitted regularly or immediately, but semi-permanent migration has become a major strategy of survival. Urban petty commodity production and wage-labouring classes have expanded dramatically. Migration is mainly of adult males; female headed households have become even more common than previously, and there is a substantial fringe of broken families with vagrant or semi-vagrant children roaming the streets of the main towns. Those few who have survived the famine with some resources intact will be in a position to dominate the assetless, at least economically, in years to come. However, the widespread breakdown of traditional redistribution mechanisms (livestock loaning, food-sharing), and of the *Shayl* (crop mortgage) credit system, indicated that almost no rural households are able to lend, and that the merchants have become unwilling to lend.[3]

A history of famine

Famine in 1985 was precipitated by drought.[4] Southern Kordofan was less badly affected than the rest of western Sudan. However, Sudan has a history of famine, traceable over the last 100 years. A review of this history will enable an analysis of the causal mechanisms which allowed the low rainfall of 1984 to have such devastating effects.

During the last 100 years there have been two major and several minor famines in Sudan. Sudan is structurally a grain surplus country; famine has been caused by negligence on the part of governments and the country's ruling and middle classes. The two major famines, 1888/9 and 1984/5, were the result of a failure to care for the food security of the people: failure to organize and encourage production, failure to store, and failure to distribute to the destitute. In 1888/9 the legitimacy of the Mahdist state and cause was not undermined in the eyes of the rulers by the existence of famine; in 1984/5 the fact of famine was suppressed by a dictator and his sycophants desperate to hang on to power. Neither regime drew its legitimacy sufficiently from the well-being of its people.

In the intervening 100 years or so, famine on a large scale was avoided, at first because the Condominium administration *was* concerned about preventing hunger. The absence of hunger was a source of its legitimacy in the eyes of the world, and especially of Egypt, its partner in Condominium. Later, famine was avoided because of the development of capitalism and a national labour market (O'Brien 1980). The British colonial state set up and later implemented a set of Famine Regulations, but the administrative procedures for dealing with widespread hunger or the threat of it have lapsed since the early 1960s. The gradual development of a national labour market in which labourers moved from poor, famine-prone regions to regions where capital was being accumulated and wage-labour opportunities were generated, in effect gradually substituted for the Famine Regulations which were designed for a situation in which the labour market remained highly segmented. However, the national labour market proved quite inadequate as an insurance mechanism in 1984/5, when the extent of crop failure greatly reduced work opportunities and wages.

Available sources on 'The Great Famine' of 1888/9 are generally anti-Mahdist. The following paragraphs should be read with that in mind.

There had been famine in 1885, due to the neglect of cultivation during the war of independence against the Turko-Egyptian regime. However, a severe drought in 1888 precipitated the country's first recorded major famine. Since the Sudan had been at war either internally or with Ethiopia or Egypt since 1881, it was hard to distinguish the effects of war from those of drought. The Rinderpest epidemic may also have been a contributory factor.

Ohrwalder (1895) gives a graphic description of the distress caused. Large price rises (from $12 to $60 per *ardeb* of *dura* in Omdurman; up to $100 in Dongola and Berber, and $250 in Kassala and Gallabat) led to much starvation and 'thousands of strange episodes' concerning orphans and beggars. Many areas of the country were depopulated: the Shukriya tribe declined from 40,000 to 4,000; a military force at Gallabat was reduced from 87,000 to 10,000. Karkoj and Sennar, the 'granaries of the Sudan' were 'desolated by famine' (ibid.: 315). There was a lot of fighting in western Sudan in 1888/9 which could not have helped matters there (Holt 1958). Livestock were eaten, and there was massive movement of population. Ga'aliin from the Northern Province settled in the Gezira at this time (Barnett 1977: 61). Omdurman did not starve but only because grain was brought from the Shilluk at Fashoda in the south, at that time an independent kingdom. This grain is alleged to have been sold at $6 per *ardeb* to the Khalifa's Baggara followers and $60 to all others in Omdurman.

The government's policy, beyond favouring the Baggara generally and the Ta'isha (the Khalifa's people who had recently been brought to Omdurman from Darfur at great expense) in particular, was to requisition sufficient food for the capital at all costs (Holt 1958: 175). Not only had the Ta'isha migrated there, but the Khalifa had also ordered the population of the Gezira to Omdurman and destroyed many of their villages (Ohrwalder 1895: 306). The Gezira was later raided for grain (Holt 1958: 175–6). Grain had previously been requisitioned from householders in Omdurman. The concern of the regime was primarily to support itself and its 'overgrown military caste' (ibid.: 174). It is not surprising that there was large-scale migration to the capital from the provinces to obtain food. It was these refugees who suffered most rather than Omdurman's trading-based indigenous communities.

It is difficult at this distance and with such a paucity of information and bias of contemporary observers to judge this episode fairly. It may be that such a policy was inevitable in the early years of a regime threatened from outside and quite insecure. After the famine there was certainly the intention to rehabilitate agriculture. At all times during this period it was probably difficult to restrain the military machine – the 'hungry Ansar' – from extracting what they wanted from cultivators.

The Mahdist regime clearly depended on an extractive and arbitrary military administration. Faced with famine, the ultimate threat to survival, its response was to secure its own narrow interests. It did not have the capacity to do otherwise without the danger of loss of recently won nationhood. There was little concept of public welfare.

By the time of the reconquest in 1898, British colonialism had established in India a method of coping with famine. The essential components in the treatment of famine were: provision of employment for the able-bodied, food distribution for the destitute through 'poorhouses' and through purchase, and tax remission. Loans for consumption or seeds were sometimes also given. More positively, famines eventually provided the major spur to agricultural development and the state's key role in improving agricultural education, land reforms, rural credit and the construction of an irrigation infrastructure.

This was the intellectual and policy background against which the British administration of the Sudan saw famine when it occurred in 1914/15. Politically, the administration justified its own existence partly on the grounds that it would not tolerate the famine conditions prevalent before the reconquest. Later, after Egypt's independence in 1922, Egypt was always looking over the British shoulder: 'Every useful work by Egypt in the Sudan will benefit her and strengthens her rights in the Sudan and renders them irrefutable',[5] ran a comment on a file in 1932 in reaction to Egyptian financial aid to famine victims in Fung province. The 'food weapon' was used even then. News of the famine had appeared in an Egyptian newspaper *Hadarat el Sudan*, and Prince Omer Tessoun had sent £E100 to the newspaper as a donation for famine relief.

In 1915 some districts of the provinces of Dongola, Blue Nile, and White Nile were struck by famine. The government purchased and distributed *dura* from India. In Dongola the problems were one of a low flood of the Nile and the loss of oxen, which reduced the cultivated area dramatically. Relief works, employing hundreds of people, were opened to remodel existing basins and open up some new minor ones for cultivation along the Nile, and for other infrastructural works to improve agriculture. Temporary hospitals and poor houses were opened, and tax remissions were allowed where necessary. Seed loans were given.[6] Elsewhere rains failed, and overall about 200,000 fewer *feddans* were cultivated in 1913/14 as compared to the previous year. The effect of this was seen in the high *dura* prices of 1914 (Beshai 1976: 290). There were three years of drought from 1911 to 1913. In 1914 *dura* prices in El Obeid were five or six times what they were in 1910 (ibid.: 308, n.17). Apparently, the rest of the war period saw good *dura* crops with record amounts of exportable surplus (ibid.:290). From 1910 to 1914 were famine years in the sultanate of Darfur, undoubtedly helping to undermine the sultan's rule, and pave the way for the British conquest in 1916 (de Waal *et al.* 1986: 9).

After the war, national famine regulations were drawn up. Although they were not applied to all provinces immediately or equally, they provided a guide for government action in case of famine for the next thirty years.

Since these regulations represent the outcome of the most serious governmental thinking on the subject which the Sudan has ever seen, a substantial extract is given in the Appendix to this chapter (pp. 66–71). The components of governmental reaction included most of what today's disaster management industry would prescribe. There was an early warning system, taking account of signals indicating famine; the only modern addition to this list would be the use of satellite imagery to measure vegetation growth over wide areas. There was a system of preliminary measures to make sure government was prepared for scarcity and a system of 'test works' designed to see whether people came forward for public works at very low wages and hard conditions. Taxes were to be remitted at this stage in order to 'reassure the cultivating classes'. Finally, once the governor-general had decided that famine conditions pertained, relief works were to be opened to provide employment and food imported into the area to be distributed to the needy. Guidelines for the use of private charity were also laid down. Beyond these specifics, the general obligation of the administration was recognized: 'when any officer finds any person in urgent need of relief ... he must take the responsibility of supplying the relief at once and then immediately report his action for sanction'.[7]

The means to bring relief had to be sought by affected districts from central government, in particular the civil and financial secretaries. Provinces also had 'Grain Currency Accounts' which could be used to import grain. A reading of the 'famine files' in the UK National Records Office shows that government response was generally fast, communication by telegraph where needed, with a rapid decision-making process between civil and financial secretaries on matters of urgency. Where there were delays, there were also complaints, as for example during the depression years when central government tried to avoid spending additional amounts on famine relief.[8]

The severe years of famine were 1932, 1938, and 1949. The 1932–4 famines were largely a result of reduced incomes caused by low crop and animal prices, sometimes combined with natural phenomena, especially locusts. Low prices also undermined the government's tax base, on which famine relief depended. The major areas affected were the main non-irrigated cash-crop producing areas (Kordofan, Fung province) and the southern provinces, where localized famine has been endemic throughout the century. Darfur was not mentioned in the files at this time. Nor were Kassala and the Red Sea provinces. Perhaps these were not involved in the market to the same extent, and were insulated from the main cause of these famines. Perhaps Darfur was beyond the administrative pale, further from Khartoum than administrative and political eyes could see. Thus the governor wrote to the civil secretary in November 1939 after a year of severe locust attacks, remarking that the problem was the distance of Darfur from the El Obeid rail head if there was a shortage: 'The people of Darfur are used to tightening their belts and living on wild rice in the South and *haskanit* and other grain seeds in the Centre and North and to some extent this is being done this year'.[9]

Table 3.7 indicates how severe the loss of income in some rain-fed areas was during the depression of 1930–2. These facts add a new dimension to the conventional picture of the Sudanese economy during the depression (Beshai 1976: 291–5). It was much worse affected than previously thought from analysis mainly of the irrigated sector.

Table 3.7   Indicators of loss of income in eastern Kordofan, 1930–2

| | Value (£E) of gum arabic sold at: | | Grain and sesame exports (tons) |
	Um Ruwaba	Er Rahad	
1926/7	12,153		11,151
1927/8	12,967		19,113
1928/9	13,876		18,506
1929/30	38,042	35,876	25,252
1930/1	19,498	12,970	11,951
1931/2	8,234	6,453	8,065

Source: NRO Civ. Sec. 19/1/2, Letter from district commissioner, eastern Kordofan, to governor, Kordofan, 13 September 1932.

Broadly speaking, the same areas, with the addition of Darfur and Red Sea suffered in 1939, but to a much lesser extent, if we judge by the amounts of grain distributed. As war approached, food became a matter of military concern since the country would have to support an increased military presence and, if possible, repeat the exports achieved in the First World War.

After the war, 1948/9 was a year of rain failure in northern areas. Its significance lies in the interest shown by the political parties, Egypt, and the budding free press in drawing attention to famine or contributing to its alleviation. With independence on the agenda, this was very much a political famine in its treatment. It affected nomads especially severely, since areas north of latitude 15° were largely occupied by nomadic tribes.

The actions taken to prevent famine are an indication of how routine famine administration became. In some cases it seemed to be just another budget which could be tapped for public works: one wonders whether there was always a genuine threat of famine, or whether it was not an excuse to get roads built or maintained. This may have applied especially in the south and clay areas of the north, where roads were critical to the administration, and the famine budget provided an opportunity to build roads. However, not all administrators looked favourably upon famine relief. The governor of the province of Blue Nile wrote to the civil secretary in 1939 about the Ora and Jum Jum tribes of the Ingessana Hills who were short of food:

I am loath to start any form of relief work before it proves essential because:
(a) The people of this area are already improvident and any form of relief work will only increase their natural inclinations.
(b) People eat roots and beer. Beer is absent but roots are available. Without medical advice I cannot be sure the people are below par.
(c) Once relief works are started demands for it will snowball.
(d) The 1932 famine papers show beyond doubt that real famine conditions did not exist there at any time and that where poor-houses were provided they were not used.
(e) There is no work they can do. The Military would presumably veto road construction. Relief works would not add any real benefit to the area.[10]

Frequently, administrators were loath to undermine merchants by the sale or distribution of subsidized grain: the merchants were generally seen as allies in the prevention of famine, as they provided relief themselves on occasion. At other times, however, merchants were blamed for exporting food crops from areas of scarcity, debts to them were cancelled by the administration or leniency shown in the enforcement of debt repayment. Administrators sometimes felt they had to check that merchants had adequate supplies, for example, to tide an area over the rains.

Support for the merchants was because there was no practical alternative to them in most of the country under the financial and administrative constraints of colonial government, not because they were or represented the ideological holy cow of the free market as in nineteenth-century India. British colonial rule in the twentieth century, by contrast, tended to show a certain disrespect for the institutions of the free market.

On the other hand, the government did try to maintain a distinction between relief through public works and charity. The latter was ideally to be restricted to the aged, infirm or destitute. Able-bodied people should always work for their subsistence.

With hindsight one could say that the entire famine administration was erected to prevent localized famines in an era when people could not easily move out of a famine-stricken area. There was an ambivalent policy towards the building of local reserves, and a very hostile attitude towards the erection of internal trade barriers, which might have been a means of creating local food security. Generally, free trade was not to be interfered with. In January 1939 a report from the Kordofan governor anticipated a shortage of grain in eastern Kordofan and Tegali, which normally depended on grain from eastern Kordofan. Production of grain in 1938/9 was estimated at 66,000 *ardeb* by comparison with 86,500 in 1937. However, in 1937/8, 77,000 *ardeb* had been exported by railway, leading to a local shortage during the rains of 1938. 'I do not know if it is possible but what seems to be wanted now is to forbid the export of grain from East Kordofan railway stations.'[11] The civil secretary's response was that an internal trade barrier would be 'an extra-ordinary measure needing extremely strong reasons to justify it'. If necessary, the area should export and then re-import.[12] The financial secretary commented that there was no law under which the internal movement of grain could be forbidden. And in any case, the proposal was not wholly sound, as grain could be carried by camel.

He inferred that there should be no difficulty in supplying areas which were temporarily short of food.[13] Presumably, the fact of the matter was also that eastern Kordofan, a major cash-crop and food-exporting area, supplied not only critical export items but also food for the capital city.

Policy on local reserves was more liberal. Remote areas were certainly allowed to maintain modest reserves. In Equatoria[14] and Darfur[15] chiefs were encouraged (in Darfur with loans against their salaries) to keep and distribute reserves. In 1932 a system of grain reserves started in the Blue Nile under village elders, was said to be flourishing in the early 1930s;[16] in the southern provinces each area was supposed to have a reserve.[17] These efforts constituted interventions in the market – politically more practical, perhaps, than trade bans which would anger traders.

National wildfood reserves, or forests, were of course generally accessible then, and people relied heavily on these during periods of scarcity. Government was even prepared to allow foragers into the Dinder game reserve in 1932 when the district commissioner lodged a complaint that people had not been permitted.[18]

Exports of grain from the country were normally controlled under the Contraband Goods Ordinance, that is as a normal function of the Customs Department. The government itself was sceptical, however, about its ability to control exports to Eritrea, Abyssinia, or Chad. A monthly average of 5,000 tons had been exported between 1936 and 1939. Occasionally, there were government-to-government negotiations about exports: in 1931 the French administration in Adre, Chad, requested permission to purchase grain in Darfur.[19] In October 1939 the Italian government of Eritrea requested a monthly export of 1,000 tons of *dura* from Sudan, which was granted, despite worries of a shortage.[20]

In the Second World War, government control of both exports and imports of food and of internal trade received a boost. Internal price control was imposed in 1941 but was not successful, according to Beshai, until 1943, when

> a Dura Purchasing Commission was established to inspect producers' supplies, purchase all surpluses, and ensure fair allocation to different districts. The Commission worked with authorised dealers only, but the real factor which secured its success was strict government control over transport facilities. 80 thousand tons were purchased by the government between October 1943 and March 1944.
>
> (Beshai 1976: 298)

Internal prices remained low by comparison with other Middle East countries.

During the war *dura* and *dukhn* were substituted for cotton cultivation in the Nuba Mountains, the Tokar Delta, and Torit district, Equatoria (Badal 1983: 26) but not in the Gezira, as the Sudan Plantations Syndicate was strongly opposed to it. This broadening of agricultural policy concern from cotton to the cultivation of other crops was partly caused by the need to make good deficiencies in Egyptian food crops, as well as the internal supply and security needs of the Sudan. A substantial effort went into increasing wheat production along the Nile north of Khartoum, though it was recognized that the British Middle East Supply Centre could only look for a small import-substituting contribution from the Sudan as far as increased wheat production was concerned.[21] Elsewhere, too, there was a concern to promote 'alternative crops': in Darfur a reaction to the famine of 1939 was to promote cassava, irrigated wheat, potatoes, sweet potatoes and *dura 'feterita'*.[22] However, famine conditions in Sudan did not lead to the level of investment and development by the state which occurred in India. Only in the 1950s was serious non-cotton agricultural development inaugurated.

The Emergency Supplies Committee decided in August 1939 to create and maintain a government *dura* reserve – 'to reassure the population'. The primary functions of such a reserve would be to check any rise in price of *dura* to the hardship level by retail sale to the public at a fixed price at certain centres and by distribution in centres where supplies had dried up by failure of crop, locust scare, market manipulation, or other causes. It is not known how much was in this reserve, when it began operation, nor when it ended. Apparently there was still a reserve kept in Khartoum North until the early 1960s.

What was the colonial motivation for attempts to prevent famine? The colonial background in India has been mentioned: famine prevention was on the agenda of the British Empire. Second, the British could not afford to repeat the Mahdist failure on food security if it was to retain any legitimacy, and legitimacy it needed to gain the co-operation of Sudanese citizens for the inexpensive and effective administration of the country.

Egypt's participation in the Condominium was a powerful factor. As independence approached, and Egypt sought to regain lost influence through political alignment with pro-Egyptian forces in Sudan, its impact was enhanced. In 1949, in addition to gifts of grain, Egypt suggested that a UK Parliamentary Commission be established to investigate food needs.[23] The government of Sudan was admonished by the Foreign Office for its sensitivity on this issue.[24] Within Sudan famine was, understandably, a political issue. The newspapers, most identified clearly with one or another political party, were full of it. As much as potential embarrassment to Egypt, the government was aware of Sudanese middle-class reaction to famine. There were significant private donations to the Fung province famine of 1932.

In 1949 pressure on the government came from explicitly political organizations, particularly their newspapers, which were full of advice on how to avert a repetition of 1948. Sayed Abdel Rahman El Mahdi presided over the collection of £3,562,020 for famine relief in northern and eastern Sudan. The list of twenty-five contributors was published in the local press.[25]

In the First and Second World Wars military security was added as a powerful motivating force: it may also have been a critical motivation for the formulation of the original Dongola Famine Code. Outside war, security was a local factor which prompted administrative action, especially in the southern provinces during the 1930s.

The Upper Nile governor wrote in 1939:

> I am especially anxious to do something for the Western Nuer on administrative grounds as well. Their country is so hopeless and when their meagre crops fail they are liable to become restless and a Dinka raid is always uppermost in their thoughts.[26]

On occasion, as at Jebel Tuleishi, famine relief helped in the process of pacification. There had been a bad harvest and locust attacks in 1931/2, and merchants had bought up stocks 'for sale in other Nuba areas where conditions at that time were far worse'. Famine relief made the Jebel Tuleishi 'friendly to the government for the first time'.[27]

The years between 1950 and 1970 were by and large years of plenty. Rainfall was reasonable over most of the country until the late 1960s. The Korean war created an economic boom in the early 1950s and the cash economy swelled. Investments in irrigation, roads, water supplies and education created jobs in government service, and the expansion of irrigated and mechanized agriculture opened up the labour market (O'Brien 1980).

The nature of the growth of the labour market was particularly significant. As capitalism expanded, more and more temporary employment in harvesting cotton and *dura* was available. This employment occurred largely from December onwards and was thus, to an extent, complementary with peasant farming. Peasants could harvest their crops and then migrate to find work and remit or take money home. This had long been done in gum arabic collection, and the practice was extended to other crops and other areas. If crops were bad or animals died, members of the family could now compensate for this with earnings from wage-labour. The latter were never large but contributed to the ability of peasant households to withstand famine conditions in their home areas.

Generally speaking, it became an accepted strategy for deficit householders to eke out their incomes by seasonal migration. The economy of the core riverain areas depends to this day on migrant labour: this is undoubtedly the major explanation for the lack of serious enthusiasm for development in the peripheral areas shown by both the colonial and post-colonial state (Ibrahim 1984), and explains why famines did not provoke development efforts. Given that framework of limited and uneven development, migration had undoubtedly prevented famine until 1984/5. It rendered famine relief, on the small scale it had been practised by the British administration, less necessary.

The southern provinces participated very little in the national labour market, partly because of the colonial 'Southern Policy' of separate development and partly because of the civil war (1955–72). They were the only areas to suffer famine conditions during the 1950s and 1960s. Shipments of grain from northern Sudan were frequently required.

The Sahel drought of 1968–73 left Sudan comparatively unscathed. Famine was averted because of the strength and continuity of the developed core capitalist economy which could provide labouring opportunities, the existence of a large mechanized *dura* sector in Gedaref district and Blue Nile, and substantial irrigated *dura* production which could offset low yields in the rain-fed sector. With the exception of 1968/9, when irrigated *dura* yields were low, and the rain-fed area planted in Gedaref and Blue Nile was halved by drought, overall production of *dura* was maintained during this period.

The biggest population movement as a result of this drought was that of the Zaghawa in Darfur. Thousands moved south from their millet fields and grazing grounds in northern Darfur to the sandy areas of southern Darfur where they became peasant farmers or labourers. Others became traders in Nyala and elsewhere (Abu Sinn 1980). Hales has claimed that 30 per cent to 40 per cent of the population of the Northern District in northern Darfur migrated southwards during the 1970s (Hales 1979). Ibrahim has argued that cultivators generally react to a dry year by increasing the area cultivated (Ibrahim 1984). The more labour available, the greater the expansion, resulting in exposing wide areas of soil to wind erosion in the most arid seasons. This is both a survival strategy and a commercial one, since grain prices will be high in a dry year. If a family cannot expand its farm to counteract the reduced yields of a drought year, it will most likely migrate. Similarly, nomad migrants were typically those with small herds, often of cattle and goats, which withstand drought and thirst less well than camels or sheep.

Hunger there undoubtedly was: whether it became starvation or not and how widespread it was, was not researched. According to Yusuf Takana, at the time administrative officer of Dar Masalit, the Gimr tribe in the Kulbus area of northern Dar Masalit were hard hit. Many migrated to El Geneina and Kerenik. The government transferred some grain to these centres and to Kutum, to be distributed to the destitute and, otherwise, sold at subsidized prices. No effort was made to distribute the grain in the rural areas, and Takana's own efforts to do so in Kulbus village were penalized with transfer to Khartoum. Takana commented that up to 1960 all districts in Darfur had underground silos (*shuna*) and funds were allocated to the councils to enable the administrative officers to fill the stores. Up till 1971 there was a large amount stored in El Geneina. These systems in the 1970s collapsed for lack of funds and an absence of concern with famine-prevention policy.

The experience of this major, if not devastating, drought did not lead to the institution of a new food security system. Both 1970/1 and 1971/2 were years of bumper harvests in eastern and central Sudan. Subsequent years were less good but quite reasonable. Despite this, only in 1967/8 and 1970/1 were large purchases made on the market for long-term storage in the new Gedaref silo. Long-term storage capacity in silos remained at 150,000 tonnes (100,000 tonnes at Gedaref, 50,000 tonnes at Port Sudan) from 1967 onwards.

The idea of a national reserve came back into policy discussion in 1976. Previously, long-term storage had been for price maintenance and export purposes. On occasion there was clear top political involvement in the management of the silos. In 1970/1, a year when private investors were scared by the spate of nationalizations, Zein El Abdin, then vice-president, visited Gedaref at harvest time, found that prices were very low and, with his eye on the forthcoming presidential election which could be influenced by the Gedaref farmers, ordered the Agricultural Bank of Sudan to purchase a large quantity.

In 1981 political involvement led to extreme mismanagement. The

Gedaref silo was ordered not to take any private storage as the minister of finance had ordered the Bank to buy 100,000 tons for the national reserve. Merchants also wanted to use the silos for storage, anticipating a price slump and the need to export. In the event, the Bank only filled the silo to 50 per cent of its capacity in April, its budget not being released till most of the crop was harvested and stored privately. As a result, merchants obtained permission from the provisional commissioner of Kassala to erect large private stores in the crop market in Gedaref.

The year 1981/2 was better than 1980/1, and this time the Bank brought 50,000 tonnes across the country at £S15 per 90kg sack, a total bill of £S8.5 million. This was the last good crop year before the disaster of 1984/5. The national reserve was extensively used in 1983 for the first time (Table 3.8) (previously, consignments of a few thousand sacks had regularly been sent to the south and occasionally to the west). Because of this the Bank purchased 30,000 tonnes in the poor crop year of 1983/4 at four times the price of 1981/2. The bill was £S19 million (see Table 3.15). Some of this was disgorged during 1984 (Table 3.9), leaving a balance of 67,627 tonnes at the end of 1984.[28]

Table 3.8 *Dura* purchases by Agricultural Bank of Sudan (all Sudan) 1979/80 to 1984/5

Year	Purchased (tonnes)	Average price/90kg sacks (£S)
1979/80	90,000	20
1980/1	31,000	15
1981/2	50,000	15
1982/3	2,800	30
1983/4	30,000	60

Source: Agricultural Bank of Sudan, Khartoum, November 1985.

While the formation of national-reserve policy was a step in the right direction, it was never a food security policy as such, only an idea which led to a series of *ad hoc* measures to store and distribute grain.

Nor did the drought of 1968–72 lead to significant changes in agricultural policy to benefit the drought-prone areas. Certainly, there have been a number of heavily foreign-aided rural and infrastructural projects in western and southern Sudan. But these have worked in fits and starts and have received relatively little government commitment.

Table 3.9 Use of Gedaref silo national reserve during 1983 and 1984

	Distributed to:	*No. 90kg sacks*
1983	Wad Medani	n.a.
	Atbara	n.a.
	Khartoum	n.a.
1984	Eastern Region	119,494
	Central Region	28,739
	Northern Region	14,000
	Darfur	124,296
	Kordofan	158,622
		445,150 (or 40,064 tonnes)

Source: Gedaref silo office, 26 October 1985.
Note: n.a. = figure not available.

On paper the major governmental response was the Desert Encroachment Control and Rehabilitation Programme (DECARP). Only in the early 1980s did this programme materialize into a few small, aided projects.

Unfortunately, DECARP and the prevalent and popular analysis of drought in Sudan has confused drought with desertification or environmental degradation. While there is a relationship between the two, they must be kept conceptually distinct. Drought refers to a shortage of rainfall for crop growing, livestock rearing and other means of gaining a livelihood; desertification or environmental degradation is a long-term process which may be accelerated by drought, but is also related directly to land use patterns. The emphasis till recently in policy has been on combating desertification, not on improving the capacity of communities and households to withstand drought. And yet it is drought which is the major threat to life, not desertification.

The national food security system did not markedly improve; agricultural policy did not adjust; and to make matters worse, rainfall was very erratic during the late 1970s and early 1980s. The dry phase was prolonged across much of semi-arid and savannah Sudan. Vulnerability of large sections of the population to drought increased significantly. At the same time, there were a large number of additional mouths to feed as a result of emigration of refugees from Ethiopia, Eritrea, Chad and Uganda.

Response to famine, 1984/5

For the first time in 100 years food availability was clearly looming as a problem. During the colonial period the problem had generally been one of entitlements and local food availability rather than national food availability. And the colonial political structure responded to these problems. Neither the Khalifa's regime in the 1880s nor the Nimeiri dictatorship of the 1980s was willing or able to respond effectively to the much more massive problems of food availability combined with entitlements crises. The generally unresponsive nature of the state before the April 1985 uprising is clear. However, there were limited and belated responses even then in two areas: food distribution and irrigation policy. In early 1984 three provinces – Red Sea, northern Darfur, and northern Kordofan – were clearly in need of food aid. There was pressure from the Darfur regional government on the central government, and President Nimeiri to declare a famine in Darfur. This he refused to do until he visited the region in June 1984.

> the simple, tragic, and shamefully seldom-mentioned truth is that the 500,000 people, most of them less than five years old, who aid workers in the field and the national capital now expect to die, will do so because the Nimeiri regime simply refused to react to cries for help from the regional government, USAID, and other aid organisations as early as 1983, even when the 'drought refugees', as they were called, brought their case to the capital's front-door step in mass migrations.
>
> (Ahmed 1985: 13)

Even the presence of thousands of migrants from Dar Hamid and Dar Kababish in camps west of Omdurman and up the west bank of the Nile from late 1984 onwards failed to convince the President of an emergency. Others were convinced, and did what was in their power, both through the state machinery and by voluntary efforts.

There was some belated official recognition of the critical situation: from April 1984 onwards consignments of *dura* went from the Gedaref silo to Darfur and Kordofan (see Table 3.17). In August they were joined by the Eastern Region as the major recipients of the national reserve. The regional governments paid for this grain at a somewhat lower than market rate. However, even after the disastrous rains of 1984, the central government refused to declare a national famine, as this would have undermined its internal and external credibility. In the eyes of the world and the Sudanese, this would have been another Ethiopia. Careers were at stake too. It was said that the governor of Kordofan was afraid to tell the president of his problems. The President was clearly afraid to tell God.

The second major way in which the state machinery responded was by shifting irrigated production from cotton or other crops to the staple *dura* (Table 3.10). At the last minute in 1984 the irrigation schemes managed to change their cropping pattern so that nearly half as much *dura* again was grown as in 1983. This led to a production increase of 186,000 tonnes, or 174 per cent. In 1985/6 the area of irrigated *dura* was more than twice the 1979/80–1983/4 average, and production was expected to increase by more than that, giving 414,000 tonnes additional production. In this respect Sudan has a massive advantage over other African countries: such panic measures are possible. Elsewhere they are not as the irrigation infrastructure does not exist. State control of irrigated cropping too makes it possible to accomplish a shift of this magnitude.

Table 3.10 The shift to irrigated *dura* production, 1984 and 1985 (000s *feddans*)

Area	Five-year average (1979/80 to 1983/4)	1984/5	1985/6	% change from five-year average 1984/5	1985/6
Nile Province	18	15	40	83	222
Northern Province	9	10	25	111	278
Blue Nile	55	40	64	72	116
Gezira and Managil	331	420	579	127	175
Rahad	25	70	90	280	360
Suki	15	21	36	140	240
White Nile	40	30	56	75	140
Gash	30	65	75	217	250
New Halfa	40	60	145	150	363
Tokar	19	35	58	184	305
Total/average % change	582	766	1168	144	245

Source: Government of Sudan, Ministry of Agriculture and Natural Resources, *Crop Production 1985/6 Season: Preliminary Estimates*, Khartoum, November 1985.

The shift was not only accomplished in the late planting flush irrigation schemes, in Gash and Tokar, which by their August–October planting dates could see *dura* would be in extreme scarcity. Other schemes – Rahad, Suki, New Halfa – managed sizeable adjustments. Even in Gezira–Managil 90,000 *feddans* was shifted from wheat to *dura*. If a much greater adjustment could have been managed – say that of 1985/6 – the food deficit would almost have been halved. The challenge for the future national food security lies partly in being prepared to make such a shift on receipt of the appropriate warning signals.

However, it must be questioned as to whether the state instigated or even encouraged the substitution of 'cash crops' by *dura*. (Officials define cash crops as industrial raw material crops: in fact *dura* and *dukhn* are as often as not also cash crops from the farmer's point of view.) In the Gash Delta Scheme, pressure to grow *dura* originally came during the 1983/4 season from the Eastern Region government. This reflected pressure from the scheme's tenants. The central government agreed to allow two-thirds of the scheme (54,000 *feddans*) to be sown with *dura*, one-third remaining for castor, the scheme's main foreign exchange earning. In 1984/5 this was converted to an all-*dura* policy, again at the initiative of the regional government, which formally has no authority over the Gash Delta Corporation, and supported by the Tenants' Union. Some 48,000 acres were sown to *dura*, out of a total irrigated area of 52,000 acres. In January 1985, that is at the height of the famine, the Gash Delta Corporation was involved in negotiations with the Bank of Sudan, its financier, the Ministry of Finance and its own Ministry of Agriculture and Natural Resources. It was agreed that the Corporation could plant 20,000 acres of castor, and the Bank of Sudan gave the Corporation money for land preparation on this basis. After the Gash River flooded at the beginning of August, the new Transitional Military Council government and the Tenants' Union pressed for an all-*dura* policy, and 75,000 acres were sown.[29] This was part of an understandable over-reaction on the part of everybody in the Sudan wishing to avoid a repetition of 1984/5 at all costs. Farmers seeing the very high *dura* prices during 1984 and 1985 had every incentive to grow *dura*.

These decisions to grow *dura* was clearly taken in an *ad hoc* fashion, against pressure from the Bank of Sudan to grow a foreign exchange earning crop, even in a famine year. The balance between foreign exchange earning and self-sufficiency as policy objectives is a difficult one to strike, especially in an era when food aid is available: any state will see its interests as lying on the foreign exchange earning side, hence the pressure to grow exportable high value crops. The challenge for the future is to evolve a system and a financial infrastructure which allow this pressure to be temporarily ignored when necessary.

The struggle to grow *dura* on irrigation schemes has a long history in Sudan, not least in the Gash Delta, and this episode must be seen in that context. It is likely that much of the shift of irrigated acreage to *dura* in 1984/5 was the result of tenant agitation. In 1985/6 it was more the result of state policy, though tenant pressure was at work.

The state machinery then was able to respond in a limited way to the extreme situation of 1984/5 by allowing more *dura* to be grown in irrigated areas and by releasing stocks of grain to the affected regions. Was it able to distribute the grain? And to what extent was the state machinery prepared to 'manage' such an unprecedented disaster?

At central government level there was the Food Aid National Administration (FANA), a small organization under the Ministry of Finance, originally set up in 1968 to handle small quantities of food donated by the World Food Programme (WFP) for schools, hospitals, or as part payment for construction projects. It was not geared up to the task of distributing large amounts of grain all over the country and monitoring the distribution. By July 1985 it had received no extra manpower to deal with the emergency (Ahmed 1985: 10). As a result, once it was admitted that the government was unequal to the task, there was only a weak central agency which could supervise the international agencies involved in relief. In mid-1985, after the fall of Nimeiri, a new organization, the Relief and Rehabilitation Commission, RRC (a name with echoes of the formidable Ethiopian organization), was set up alongside the UN Emergency Office for the Sudan, to co-ordinate and supervise relief. Some staff were seconded from FANA to start the RRC. Building this organization into one capable of directing internal redistribution will be one of the elements of future relief schemes.

In the regions a variety of administrative structures were set up (Ahmed 1985: 9), but their main function in the end was to supervise the aid agencies. In late 1984 WFP managed the distribution of food aid, and worked closely with the government structure. Most of the food which was distributed before December 1984 came from government stores. In an effort to balance need with availability of grain the Ministry of Commerce slashed the estimates of need of regional governments in the west and south by half. Later USAID came to play more of a leading managerial role and eventually, after the April *coup d'état*, government was practically excluded from the arrangements. USAID's opinion was that government should have as little to do with distribution as possible. Transport was contracted to Arkel-Talab, an American–Sudanese company, and distribution to villagers was supervised and the whole process monitored by NGOs – CARE in Kordofan and Eastern Region, and Save the Children Fund (SCF) UK in Darfur. The truth is that when government had been in the picture there was very little food to distribute to many hungry mouths. Under such circumstances the government machinery was overwhelmed, and even gladly gave up responsibility to the non-government organizations (NGOs). There was no attempt to beef up the local government system to enable it to cope. This will make it difficult to build on the experience for the future.

Other aspects of preparedness were neglected too: seeds were not procured until just before the 1985 rains; neither taxes nor loans repayments were remitted; there was no attempt to buy livestock and keep them until they could be returned to their owners or to destitute nomads. The Eastern Region government did, however, put some livestock on the train to Gedaref to graze crop residues in the areas south of Gedaref. It was all that most regional governments could do to set up camps to cater for destitute migrants. These camps were, of course, sizeable commitments; for example, the population of Maaskar el Ghaba at El Obeid reached 47,000 in March 1985, and there were still 35,000 there in July. Other camps in Darfur were bigger. Perhaps the most dramatic action taken by a regional government was a control on the export of *dura* out of Gedaref, the country's major surplus area, on the grounds that there was not enough in Gedaref to feed the Eastern Region. Control was first imposed before the 1983/4 harvest, and lifted in early 1984. It was reimposed in July 1984 until April 1985, and again between July and September 1985.[30] This was the measure colonial government had refused, and it was now opposed by central government and other regional governments.[31] The ban was finally lifted in September because of pressure from merchants complaining that the price was going down too rapidly in Gedaref – a sign of the effectiveness of the control. It was only a control, not a ban, however, and a lot of grain was moved out of the region during the period (Table 3.11).

Clearly, the Eastern Region protected its own interests by this measure. Table 3.11 shows where the centres of purchasing power were during the famine, to which merchants were interested in transferring grain. Darfur barely received any. Khartoum, by contrast, received enough for half a million people to eat 500g per day. (A large proportion of Khartoum's population eats wheat not *dura*.)

The Central Region also attracted a considerable amount of grain. The control of *dura* movements was lifted in March 1985 when the

Table 3.11   Private grain trade transport of *dura* out of Gedaref during periods of controlled trade, 1984–5 (90kg sacks)

Destination	1/7/84 to 1/4/85	7/85 to 12/9/85
Eastern Region	492,413	91,501
Central Region	205,172	66,945
Northern Region	133,859	n.a.
Darfur	24,506	210
Kordofan	80,248	6,343
Khartoum	857,069	2,069
Total	1,823,267	167,068

Source: Executive officer, Gedaref District Council, December 1985.
Note: n.a. = figure not available.

merchants declared that they had 1 million 90kg sacks (i.e. 90,000 tonnes) in store.

The overall picture of government, then, is that it had a low capability for disaster management. How did the aid agencies fare? Between November 1984 and September 1985, 1,370,000 tonnes were pledged; 1,286,700 were delivered to port; 886,000 tonnes had been removed from the port, benefiting 8,790,000 Sudanese and 637,000 refugees. Assuming that only those targeted received food aid, each person would have received an average of 94kg during that period. At 0.5kg per person per day survival ration, this was enough for 188 days out of 300. In November 1984 the estimate of food deficit for 1984/5 was 800,000 tonnes. In February this had increased to 1.1 million tonnes. Another estimate was 1.7 million tonnes (Faisal Islamic Bank 1984). As in all massive relief operations it was the delays which killed. USAID, for example, the major donor, supplied 877,000 tonnes overall, of which 519,000 tonnes were under Title II, that is free, for the people of Sudan. Of the total, almost half arrived after July 1985.

The Eastern, Central, and Northern Regions were reported to be receiving enough to meet their needs by July 1985. But Kordofan and Darfur were only receiving between one-third and one-fifth of targeted quantities. The reasons were a lack of transport and the inability of Sudan Railways to move the planned amount of grain from Kosti to Nyala. The latter was due to physical constraints – lack of spare parts, track failure – and an absence of sufficient priority for food aid on the railway. The railway unions and management saw their refusal to transport food aid as a means of bringing down a regime which had sought to destroy the railways and their unions.

The transport difficulties were partly a direct result of the contract for transporting food aid being in the hands of a private, profit-seeking company. In Kordofan it was areas furthest from Kosti which did not receive their share of food aid, or only received it very late. The more food that could be off-loaded close to Kosti, the greater the profit. Presumably, USAID was not in control of Arkel-Talab's delivery schedule (Oxfam 1985b: vii, 18). As far as Darfur and the railway is concerned, Arkel-Talab had arranged with Sudan Railways that Darfur's allocation would go by rail. This was much cheaper (£S190/ tonne) than road haulage (£S370/tonne). But the 'expected' number of trains predictably never travelled, and only in June did the railways allocate a high priority to food aid. The slowness of the railways had been known since the winter to FANA, USAID, and Arkel-Talab. However only in June was a meeting held with the railways to discuss the problem. At that point, when it was already too late since the rains had began in southern Darfur, donors began to look for lorries and to airlifts to transport food to Darfur (Ahmed 1985: 14).

In Kordofan this had the result that only between 9.8 and 222g per head per day were received compared with the targets of 450g in northern Kordofan and 350g in southern Kordofan. The lowest provision was to Sodiri Area Council, followed by Kadugli, Rashad and En Nuhud. Other factors which contributed to this were the effective halving of the daily ration by doubling the period over which a particular shipment of grain was to be delivered, spillage from imperfectly sewn sacks (15–29 per cent), and bad road conditions in southern Kordofan. The daily ration *decreased* in Kordofan from March to July, in the period when the system should have been stabilizing (Oxfam 1985b: 18–19).

In Darfur the picture was far worse. The average received per day was 14g, with a range of between 2g and 34g. In addition to the gross deficit of food in the region, there were other factors: poor quality sacking, the sales of 33 per cent of the grain received in the towns, and the refusal of Darfurians to target available food on the most needy: in their view, everyone was needy. Villages followed different practices in distributing grain. Most villages probably did a blanket distribution, such that the food intended for 1 million identified as needy was distributed among 3 million (Oxfam 1985b: 39–42).

In Kordofan nomads were discriminated against even by the standards prevailing: only one-third of eighteen groups surveyed in

May/June had received any food aid (Oxfam 1985b: 19). With the exception of the destitute nomads who migrated to Omdurman in 1984 or those who ended up in camps, the famine relief effort almost completely failed to grapple with this issue. These then are some of the realities behind the plugging of the food gap by 'Reagan's *dura*' as it was popularly known.

'USAID are well ahead of other donors', commented *The Guardian* on 28 January 1985. So it remained throughout the famine. By comparison, the EEC's response, as the other potential major food donor, was slow in coming and the amount given was small. By June 1985 Canada had committed almost three-quarters as much as the EEC; the Saudi Red Crescent almost one-third. The United States achieved a major propaganda success with its food aid. It is interesting to speculate whether Sudan would have broken off diplomatic relations with United States over the CIA's and US Embassy's involvement in the export of Falasha Jews from Ethiopia via the Sudan, had it not been for the food programme. The Sudanese involved in this affair were widely publicized during the second half of 1985 and contributed to the anti-American feeling in the capital city which already existed.

A famine situation having been created by state policy, was allowed to develop into a famine, probably more so in Darfur and eastern Sudan than Kordofan, because of failures of aid agencies to react sufficiently to the crises. Only USAID responded, but the systems it set up to bypass government were flawed, and much grain was delivered too late to save lives in the most remote areas of western Sudan. In eastern Sudan the crisis in Ethiopia spilled over in the form of hundreds of thousands of families who were grouped in camps which were not sufficiently provided for in advance, despite foreknowledge of migration flows.

An interesting aspect of the famine relief effort is the entry of NGOs into the aid scene in Sudan. It would be uncharitable to say that they descended like vultures. The great majority of foreign NGOs established in Sudan, about a hundred, came first during 1985; only a few had been established and active before (e.g. Euro-Action Accord, Oxfam, ACROSS SCF (UK), and those had had very limited purviews. Most came on the grounds of relieving suffering, but were planning to move into development activities. There was thus an influx of hundreds of new development 'experts', eager to try their hand. Most NGOs received funds from multilateral and bilateral donor agencies as well as from donations. Many, particularly the US NGOs, paid quite substantial salaries to their often inexperienced expatriate and Sudanese employees. The role of NGOs has been very controversial, and would require a separate evaluation.

For the donor agencies, however, the appearance of NGOs represented a godsend – an alternative channel to the government for absorbing and managing aid money. USAID and UNHCR were most keen on this idea, and other aid agencies followed suit; indeed, it has become the policy of most western donor agencies to channel increased amounts of general development funds through NGOs during the last decade. It represented a cheaper way to disburse aid than directly employing 'experts'. And there is a fashionable belief that NGOs can achieve in development what governments cannot, because of small size, dedication, and so on. There is definitely some truth in this, though how much is a matter for debate and empirical investigation. One of the reasons that foreign NGOs are better at disbursing aid money than governments is that they, like officially aided projects, employ expatriates on much larger salaries than local counterparts. Given the personnel intensity of much relief and development work, similar results could be achieved at much lower costs by using existing, usually underemployed, government personnel. Furthermore, the use of large numbers of expatriates generates a self-preserving ideology which denies that the government personnel can do a useful job, and this tends to reinforce the donors' decision to exclude government whenever possible. All this meant that, in the establishment of a food security system the government was less able to learn from experience because it did not get it.

The influx of NGOs in such numbers is a symptom of world panic and guilt at seeing starvation on television. When NGOs were given serious responsibilities, their achievements were considerable. CARE in Kordofan and SCF (UK) in Darfur undoubtedly saved many lives by introducing and funding systems of food distribution. Both organizations received considerable co-operation from regional government. However, they were hampered in their efforts at preventing famine by USAID's transport policy: its reliance on Sudan Railways for Darfur and Arkel Talab lorry transport in Kordofan. In both cases USAID stuck to its policy for too long, and many lives were lost as a result. Interestingly, often during the famine, while SCF (UK) was withdrawing from engagement in Darfur, CARE was attempting to move into development programmes in Kordofan.

For the NGOs famine also creates a situation in which they can promote their own organization, using photographs and other visual aids to collect more money. In the next phase they must show their contributors that they have done a good job and, having relieved the emergency, are engaged in development – for the lack of development is defined by the NGOs themselves as *the* problem, *the* cause of famine. So there is a scramble for 'high profile projects' – activities which are obviously contributing to development, obviously even to the ill-informed western donor. The scramble does not apply only to NGOs of course. For example: water. If famine is caused by drought, and drought means lack of water, then obviously water is a good thing. Water is also achievable: drill a borehole, dig a well, excavate a *hafir*. This is easier, quicker, and has more immediate impact than a reforestation programme or an agricultural development programme, with all their difficulties and their uncertainties. Fresh teams of expatriates come in, unaware of and uninterested in the history of the complications about providing water in semi arid or savannah areas (Shepherd, Norris, and Watson 1987; Mohamed *et al.* 1982). Millions of dollars are going into drilling boreholes in northern Darfur as a response to drought, despite the fact that 'food rather than water was the major concern' (Oxfam 1985b: 39). Water in a semi-arid area should be provided as part of a general development programme, if it is to be provided in large quantities, since it attracts people and livestock and has serious effects on the environment. Some would argue that it is these effects, repeated on a massive scale by expansion of drinking water supplies during the 1960s and 1970s, which are an underlying cause of famine.

In such a scramble, humility before a deep problem is lacking. There is a strong belief in the capacity of human beings to control nature, and a belief in the ability of dedicated European field workers to bring about a greater degree of control, through technology or superior organizational methods.

Causes of famine

Why then was there a famine? The most popular explanation in the Sudan was that it was an act of God. Many people believe that God was angry with the government of Nimeiri. This belief was reinforced by the good rains of 1985/6. One could argue, however, that while drought may certainly be an act of God, famine does not necessarily follow from drought. Human activities intervene to determine whether or not people go hungry. A second popular explanation is that famine is a result of the underdevelopment of non-riverain regions which undermined their ability to produce. This is a general argument which might have applied at any time since the Condominium government decided to focus economic development in central Sudan, but it is nevertheless a powerful argument. A third popular and also powerful explanation is that food shortage caused famine (Sen's 'food availability decline' school).

A sudden rainfall decline is clearly beyond man's control – deforestation or industrial pollution may be partly to blame, at least for the slowly declining rainfall of the late 1960s and 1970s. The likelihood of a sudden substantial decline, at least in one part of the country if not across the whole of northern Sudan at once, must have been increasing during these years: it was predictable that rainfall would fall short of minimum needs in the near future. What was not predictable was that it would do so across such a huge area, leading to the breakdown of a large part of the rural economy and the simultaneous shrinkage of work opportunities.

The people, the government of Sudan, and the international community reacted to the long-run decline in rainfall and the likelihood of drought in a number of ways which fell short of the need. Agriculturally, farmers adapted by growing quickly maturing varieties of grain which needed less rainfall and by conserving water from *wadis* or rainfall by digging shallow trenches (*taras*) around the field. Pastoralists reacted by moving further afield with their stock, often further south, and by diversifying their sources of income (Abdalla and Holt 1981). The government and the international community responded with a series of rural development projects in western Sudan, with agricultural research a focus or a component (see Ch. 8). Most of these did not begin until the late 1970s, and were by nature long-term ventures: no short-term dramatic results were likely which could stabilize farmers' or pastoralists' production levels, or which would so increase their incomes that there was a substantial surplus to reinvest in production. The projects are not ones to which the state has been very strongly committed. It has been more committed to providing services such as water, health, and, to a lesser extent, education – vote-winning services.

The basic problem of agriculture and pastoralism in the rain-fed sector throughout the Sudan has two faces: one face is the inability of most farmers or livestock owners to retain a surplus – due to market conditions, low farmgate group prices, and high interest rates, due to the prevailing low level of technology, and due to the absence of an effective transport infrastructure. The other face is the absence of tried-and-tested technologies of land improvement in which farmers could invest if they had a surplus. The implication of lack of surplus is that little grain has been stored in recent history. *Dura* and *dukhn* are cash crops as well as subsistence crops: the pressure has been on to tilt the balance towards selling and away from storing what was not needed for consumption. This is the core of the underdevelopment of the peasant areas.

Neither this nor the sudden drop in rainfall across such a massive area, however, gives us an adequate explanation of famine in my view. Nor does the decline of food availability. Table 3.12, based on official statistics, shows that food availability had declined as low as it did in 1984/5 twice before during the last twenty years without causing famine. Per capita availability, even after exports have been deducted, shows that the five-year average 1980/1–1984/5 was higher than at any time since the early 1960s. The fourth element is the nature of the market for grain during the previous ten or fifteen years. At stake in particular is the pursuit of an export market for the country's staple food, *dura*. Export figures are given in Table 3.13.

The big increases in exports from 1979 onwards were presumably allowed to help redeem Sudan's disastrous balance of trade and payments position. In years of high production they were also allowed to prevent a slump in producer prices in the country. On occasion during the 1970s the Agricultural Bank of Sudan has itself exported *dura* through agents. In 1980 large quantities were apparently sold to merchants for export. The very large volume of exports in 1981 and 1982 was due to subsidies placed on sorghum imports from Sudan in Saudi Arabia. These were enough to raise the price to £S200 plus per tonne. However, the subsidy was removed in 1982 and Saudi Arabia imported from elsewhere. Nevertheless, prices remained high. Exports during 1979–83 were three times greater than 1975–8, while production was only marginally more (Awad 1984: 34). The alternative policy of surplus storage has been on the agenda, not least as presented in the 1983 report of the government's Agricultural Production Relations Committee, chaired by Professor Mohamed Hashim Awad (ibid.: 33–4). However, it was not seriously adopted.

The private commercial banks which have flourished in Sudan since they were legalized and encouraged by government (in 1978) financed a substantial proportion of exports. Faisal Islamic Bank (FIB) was prominent among them. Its export credits, which were mostly for *dura*,

Table 3.12　Total sorghum production in Sudan and per capita food availability 1960/1–1984/5

	Production (tonnes 000s)	Production – exports (tonnes 000s)	Production (kg/person)*	Production – exports (kg/person)	Five-year averages of production per capita (less exports) (kg)
1960/1	1051	n.a.	95	n.a.	
1961/2	1434	n.a.	130	n.a.	
1962/3	1267	n.a.	115	n.a.	
1963/4	1349	n.a.	122	n.a.	
1964/5	1137	n.a.	103	n.a.	113
1965/6	1095	n.a.	99	n.a.	
1966/7	851	n.a.	60	n.a.	
1967/8	1980	n.a.	141	n.a.	
1968/9	869	n.a.	62	n.a.	
1969/70	1451	n.a.	103	n.a.	93
1970/1	1534	1502	109	107	
1971/2	1590	1535	99	99	
1972/3	1300	1206	81	75	
1973/4	1692	1603	94	89	
1974/5	1681	1636	93	90	95 (91)
1975/6	2160	2086	120	115	
1976/7	1789	1685	99	93	
1977/8	2062	2015	115	112	
1978/9	2193	2021	122	112	
1979/80	1462	1176	81	65	107 (99)
1980/1	2068	1827	115	101	
1981/2	3272	2859	181	158	
1982/3	n.a.	n.a.	n.a.	n.a.	
1983/4	1829	1573	102	87	
1984/5	1110	n.a.	61	61	114 (101)

Sources:　Government of Sudan, Ministry of Agriculture, Food, and Natural Resources, *Yearbook of Agricultural Statistics*, 1960/1 to 1984/5; Sudanow, *Sudan Yearbook 1983*, Government of Sudan: Khartoum 1983.

Notes:　*Based on population of: 1961–5, 11 million; 1966–70, 14 million; 1971–2, 16 million; 1973–81, 18 million; 1982–5, 19 million.
n.a. = figure not available.

amounted to 15 per cent of its total investment in 1982, the year of massive exports (Suliman 1984: 67–8). Its share of exports in 1983 and 1984 (as it reported it) is given in Table 3.14.

Table 3.13 Exports of *dura*, 1971–85

Year	Amount (tonnes)	Value (£S 000s)	Price (£S/tonne)
1971	32,482	1,085	33
1972	55,276	1,664	30
1973	93,953	2,922	31
1974	89,217	4,401	49
1975	45,084	2,233	49
1976	74,452	3,168	43
1977	103,834	4,767	46
1978	46,916	2,664	57
1979	172,024/196,480*	13,524	79
1980	286,249/337,511*	43,024	150
1981	241,279	42,903	177
1982	412,768	107,474	260
1983	256,162	66,555/70,759*	260
1984	24,887	7,236/7,899*	290
1985	n.a.	n.a.	n.a.

Source: Ministry of Commerce and Supply, Khartoum, 9 December 1985.
Notes: * Bank of Sudan figures quoted in Faisal Islamic Bank (1984).
      n.a. = figure not available.

Table 3.14 Value of exports of *dura*, 1983–4 (£S 000s)

Year	Exports financed by FIB	Other banks	Total	% share of FIB
1983	15,352	55,407	70,759	22
1984	2,888	5,011	7,899	37

Source: Faisal Islamic Bank 1984.

Famine, then, was a product of the way in which the Sudanese state responded or failed to respond to drought and economic crisis. In summary:

1. The state invested too late and too little in peasant rain-fed agriculture and livestock production, maintaining a low level of infrastructure and official credit, which prevented significant surpluses being accumulated within the peasant economy.
2. It failed to maintain adequate food reserves, choosing to allow exports of *dura* instead, responding to the pressure from its international bankers, farmers, and merchants.

3. It failed to shift production adequately to food crops in the irrigated sector at the critical moment.
4. It failed to admit that disaster had struck in time to mobilize international food aid.

Famine protection

This chapter has stressed the role and nature of the state in relation to the rural economy as the major cause or set of causes of famine. It was the state, under pressure from international creditors and domestic merchants, which maintained and latterly accelerated the policy of exporting grain, largely as animal feed. It was the domination of the state by merchant, officer, and intelligentsia social classes based in the riverain areas which led to the underdevelopment and resultant over-exploitation of the peripheral areas. The political system changed dramatically with the reintroduction of parliamentary democracy in 1986. At the time of writing it has been unable to resolve the major manifestation of the problem of regional underdevelopment – the renewed civil war in southern Sudan. However, the prospects for improved food security in northern Sudan have undoubtedly increased. A more measured approach to exports and the maintenance of a national reserve has been adopted, though it has failed as yet to crystallize into a solid policy. Whether the state machinery, crippled by debt massively expanded under Nimeiri and with its colonial heritage, can actually respond effectively to the development needs of famine-affected areas, remains in doubt. The constraints are the non-availability of appropriate technology; the massive dependence on aid agencies, each with its own priorities and styles of work, which prevents the adoption of rational, measured approaches, and with many aid agencies trying to bypass existing government structures either by developing their own operational capabilities (NGOs) or by persuading government to set up corporations and projects with their own administrations.

Clearly, a government and aid agencies interested in famine prevention must do several things. One is to redress the balance of mistrust, especially in medium- and long-term agricultural research for both surplus and deficit areas. This process has now started in southern Darfur and in Kordofan, in four heavily aided agricultural research or development programmes committed to improvement in risky rain-fed agriculture. A second is to evolve a national food reserve and famine policy. Again, this is now on the agenda, if not complete. A third is to ensure that aid goes to build up those parts of the state machinery which are capable of responding to the underlying causes of famine: the greater their ability to respond, the stronger will be Sudan's democracy; the stronger its democracy, the more pressure can be exerted on the state to respond to the needs of the electorate.

In all three areas there has been progress. However, at the time of writing, post-drought policy remains ill-developed. Food aid distributed to people in their villages has frozen an untenable situation. There are areas which have suffered prolonged drought and land degradation whose populations are utterly reliant on food aid. Official rehabilitation efforts are negligible. The likelihood of new mass migrations to the Nile, the capital city, and regional towns is strong if food aid is discontinued. Policy on migrant labour, wages and conditions of work needs to be reconsidered from the point of view of food security. Until there is a change in climate, it will be necessary to combine long-term investment in agricultural research with short-term rehabilitation measures (restocking, small-scale irrigation, improved local-level storage), whenever feasible, but also to find opportunities in continued crisis. Thus urban migration should not be treated as a cancer but an opportunity to develop new businesses, industries and improved urban services and housing through public works. Such an approach challenges the xenophobia of the dominant riverain classes: this is possible under democracy since the majority Umma Party stands to benefit electorally. Migration from north to south is an opportunity to develop more stable and productive systems of agriculture in the savannah areas, while leaving the semi-arid areas to those pastoralists who still have, or can acquire, livestock. The perpetuation of food aid allows too many families to continue to try to scratch a living from ecologically ruinous systems of agriculture in utterly marginal areas. The accelerated migration of men from the rural areas should attract state agencies to work much more with women, who have been socially mobilized by famines as never before. Deforestation has set the scene for extensive tree planting by the rural people themselves as a more reliable source of income in semi-arid areas than cropping. The vulnerability of millet and sorghum to drought provides an opportunity for diversification in farming systems and in the rural economy as a whole. The opportunities are legion. Whether they can be grasped depends on the evolution of democracy and the structure and capacity of the civil service and the approach adopted by powerful aid agencies towards these two issues.

Appendix to Chapter 3
Ref: Civ. Sec. 19/1/2   NATIONAL RECORDS OFFICE,
                                                              Khartoum
EXTRACTS FROM FAMINE REGULATIONS 1920

2. The following Regulations are based on the principles of Famine Relief expressly adopted by the Sudan Government.

   (a) that it is the duty of the Government to offer to the necessitous the means of relief in times of famine,

   (b) that, if a person is capable of labour, relief shall only be given to him in return for work done, and

   (c) that gratuitous relief shall be provided for those unable to labour.

3. Actual distress must be relieved at once apart from Regulations. It is all important that precautionary measures such as the remission of taxation should be announced as early as possible to put heart into the people, and in cases of actual distress relief should be given at the earliest possible moment. When people are on the verge of starvation a day or two's delay in giving relief may reduce them so much in condition that recovery is hopeless or protracted.

4. These Regulations fall under three heads:

   (i) Permanent or annual provisions for discovering and dealing with famine if it should occur.

   (ii) Provisions dealing with threat of famine.

   (iii) Measures of relief when famine actually exists.

Part I
System of intelligence

5. (i) Comparative statistics and reports for each district with reference to the following matters should be useful in indicating the prosperity of the areas or the reverse:

   Nile Gauge
   Rainfall
   Condition of crops and agricultural stock
   Market price of foodstuffs and cattle
   Death rate and prevalence of disease

   (ii) Programmes must make provision for the employment for 6 months on work within the district of a sufficient proportion of the population.

   (iii) The Governor will decide what is a sufficient proportion to be provided for, but in districts liable to famine this should not be less than 20 per cent of the population.

   (iv) In considering the programmes the Civil Secretary will arrange as far as possible with the Governors and Heads of Departments concerned that a sufficient number of Government officials and employees or other persons who may be willing to offer their services will become immediately available for carrying out the Relief Programme in any district where famine brings it into force.

(v) In submitting his programme the Governor shall state the extra staff and number of tools or implements beyond the Province resources which the schemes will require.

Part II
Preliminary measures of preparation and tests when scarcity is imminent

7. *Premonitory symptoms*
Apart from the failure of the rainfall or the Nile and the movements of prices, the following premonitory symptoms are generally observed – the contraction of private charity indicated by the wandering of paupers, the contraction of credit, feverish activity in the grain trade, restlessness shown in the increase of crime, unusual movements of flocks and herds in search of a livelihood.

But it should be remembered that in the Sudan, where the population is small and scattered, it will be safer to rely on personal inspection of the people themselves than on any symptoms or tests.

8. When anxiety is felt as to the imminence of famine it is of the greatest importance to make early and liberal preparations in advance and thus put heart into the people. There is no greater evil than the depression, and consequent physical deterioration, of the people. Once the preparations have been made, it is necessary to wait on events, on the one hand carefully observing the symptoms of distress and on the other guarding against false alarms. In this matter it is necessary to take a certain amount of financial risk. The money spent in preparations may indeed be wasted; but the loss is trifling in comparison with the expenditure which want of preparation entails.

The measures to be taken are of three kinds:
(i) Preparation to be made in advance.
(ii) Measures to encourage the people.
(iii) Tests as to the reality of the distress.

(i) *Preparation*
(a) The governor will arrange for personal inspection to verify reports and symptoms of scarcity or distress.
(b) The Governor in consultation with the local representative, if any, of the Medical, Public Works or Member of the Egyptian Irrigation Service will revise the programme for relief works, select the works to be opened first, arrange for tools and plant to be in readiness for immediate use, select sites for camps and make arrangements for strengthening the staff.
(c) He will organise village inspection for discovering cases of actual want.
(d) He will organise non-official relief and private charity if any is available.

(e)   He will open or arrange for immediately opening when required poor houses and kitchens for gratuitous relief.

(ii)   *Encouragement of the people*
Where possible it is of the greatest importance to give liberal assistance at this stage to agriculturists in the constructions and repair of *sakias* and any other agricultural implements, seed and the like; by this means it may be possible to save some of the crops, while the assistance is also useful both for employing labour and to give confidence and morale to the people. Assistance should be in kind and money advances are to be avoided.

The prompt announcement of any remissions and suspensions of land revenue which it is decided to grant on account of the failure of crops also does much to re-assure the cultivating classes.

(iii)   *Test Works*
When the Governor cannot be sure that there is actual famine he may open test works, reporting in the same manner as directed in Section II.

For this purpose he will select one or more of the relief works in his approved programme and will start them on the lines proposed subject to the following considerations which distinguish them from relief works proper.

The object of test works is not to relieve famine but to test the presence of it; not to appease hunger but to find out whether people are hungry. Therefore:

(a)   The conditions of labour must be hard but not repellent.
(b)   The tasks should be slightly heavier than on Relief Works.
(c)   The payment should be strictly by result.
(d)   No allowance should be made to dependants.
(e)   Women and children who work may be given cooked food or an allowance of *dura* instead of cash wages, if there is reason to believe that they are not really in need of relief.

9.   When the Governor finds reason to believe that actual famine is imminent, he shall report through the Civil Secretary to the Governor-General giving detailed reasons and statistics and a statement of the Medical and Railways Departments mentioned hereafter.

He shall also organise relief inspection.

He shall also, in consultation with the Medical Officer of Health, forward to the Director, Medical Department, Sudan Government, an estimate of the additional medical staff as well as of equipment, medicines and special diets required.

He shall also give notice to the General Manager, Sudan Government Railways and Steamers, or his representative, of any requirements for the immediate transport of foodstuff, grains or fodder

Part III
Famine

10. When the Governor is of the opinion that famine exists in a District he shall report all the facts to the Civil Secretary and it shall be for the Governor-General (or the Governor-General in Council) to decide whether a District has passed the stage of observation and test and to declare when famine exists in it. This decision will be communicated at once to the Members of Council.

11. When famine conditions have been declared to exist in a District, immediate steps shall be taken for the commencement of relief operations in accordance with the approved programme, but the Financial Secretary, Sudan Government, must be kept fully informed as early as possible of the liabilities involved.

    The first steps shall ordinarily be:
    (a) To convert test works into relief works by the modification of tasks, etc.
    (b) To open relief work where required.
    (c) To commence distribution of gratuitous relief subject to personal inspection.
    (d) To open poor houses.
    (e) to import grain and fodder.

12. *Relief inspections*
    The objects of Relief Inspection are twofold:
    (a) To ascertain by personal inquiry the real condition of the people, to judge to what extent the measures of relief introduced are effecting their purpose and to see that all who require relief in any of the approved forms are able to get it.
    (b) To provide an effective machinery for the distribution of gratuitous relief at the homes of the people and for the supervision of village work. The public servants to whom the duty of relief inspection is allotted shall, as far as practicable, daily visit every house in the village to discover cases of destitution, and shall deal with such cases in the following manner:
    (a) Any person able to work shall be made to go to the nearest relief work, being given, if necessary, an order of admission.
    (b) Any person entitled to relief in the village shall be entered on the village gratuitous relief register.
    (c) Any starving wanderer found within the village boundaries shall, if able to go, be provided with an order of admission to the nearest poor house and if necessary shall be given food for his journey, or, if unable to go, shall be placed temporarily on the village gratuitous relief register.

13. Relief operations fall under two heads:
    Relief Works when relief is only given for work done.
    Gratuitous Relief in villages, kitchens or poor houses.

14. *Relief works*

In organizing and managing relief works, the following general principles should be borne in mind.

(i)   The system should be one of pure payment by result. No minimum wage is fixed but only a maximum which, subject to an addition in the case of a ganger or rais, should be the lowest amount that will maintain healthy persons in health.

The task by which maximum wage is earned in the case of ordinary adult labourers should be ¾ of the amount of work commonly performed by able-bodied labourers in ordinary times. If the worker performs 95 per cent of the full task he should get the full wage. Work short of this gets proportional short payment, and no work gets no wage.

(ii)   Special tasks should be arranged for the weakly and they should be formed into special gangs except when it is not desirable to separate them from their natural protectors.

(iii)   When the works are distant from the house of any of the workers, gratuitous relief must be given on the works to dependants of the workers, when such dependants are too young or too infirm to work.

(iv)   Care should be taken to avoid getting the works overcrowded either by dependants who might be provided for in their villages or in poor houses, or by workers who came for the wage before they have come to the end of their own resources.

(v)   It is very important that a sufficient number of relief works should be open, or in readiness for opening, to employ all who apply.

(vi)   As soon as people come to the work they should be classified and put to work at once. If for any reason they cannot be put to work at once they must be given the dependant's allowance until they are.

(vii)   The works establishment must be well organised and strong enough to keep the people up to their work.

Proper arrangements must be made for water supply and sanitation. There should be frequent inspection by the administrative as well as by the expert officials.

15. *Gratuitous Relief*

A.   By distribution of food to persons continuing to live in their own houses. Such distribution should usually be made from a cookhouse or kitchen.

B.   Poor houses, in which the persons receiving relief are detained subject to a certain amount of discipline.

In both cases the recipients should be given if possible such light work as they may be able to do.

Part IV
Miscellaneous

16. One important point to be borne in mind by officers of all grades is that irrespective of these Regulations, immediate relief must be given to any person found actually starving. Therefore, when any officer finds any person in urgent need of relief which under these Regulations he cannot sanction, he must take the responsibility of supplying the relief at once and then immediately report his action for sanction.

17. When charitable funds are subscribed by the public and entrusted to Government for the relief of distress, the following objects are recommended:
    (i) Supplementing the recognized state relief, e.g.:
       (a) by gifts of blankets and clothes.
       (b) by gifts of extra or special food or medical comforts to the aged or infirm persons of respectable birth, patients in hospital, children and the like.
    (ii) Maintaining or subsidising shops for the sale of grain at cheap rates to selected persons in reduced circumstances.
    (iii) Assisting with grants of money:
       (a) Agriculturists in want of seed grain, bullocks or fodder, or implements of husbandry, or who require support in the interval between sowing and harvest.
       (b) Artisans who have lost their tools or stock in trade or are otherwise in reduced circumstances.
       (c) Persons leaving state relief to resume their ordinary avocations.

18. Proper records and accounts will be kept of all receipts and expenditure and of all loans or advances in money or kind with a view to possible recovery.

Notes and references

1. UNEOS Sudan monthly situation report, September 1985, Khartoum, 7 October 1985.
2. UNICEF report on the Kordofan drought and the Masskar el Ghaba Camp (El Obeid), El Obeid, 19 December 1984.
3. The labour market is thus a critical prop for rural families in hard times. If that labour market collapses, poor families will be quickly thrown back on their own inadequate resources.
4. Rainfall in northern Kordofan and northern Darfur in 1984 varied from 4 per cent to 51 per cent of the 1951–80 averages.
5. National Records Office (NRO), Khartoum Civ. Sec. 19/1/3, p. 19; Egyptian press résumé.
6. NRO reports 2/8/29, memorandum by General Sir Reginald Wingate on the finances, administration, and condition of Sudan, 1914 (pp. 2 ff.).

7. NRO Civ. Sec. 19/1/2, extracts from Famine Regulations, 1920.
8. NRO Civ. Sec. 19/1/1, correspondence between governor of Fung Province and the civil and provincial secretaries, 1932.
9. NRO Civ. Sec. 19/1/2, p. 1336, letter from governor of Darfur to civil secretary.
10. NRO Civ. Sec. 19/1/2, pp. 1301–2, letter from governor of Blue Nile to civil secretary, 5–3–39.
11. NRO Civ. Sec. 19/1/2, p. 1293, letter from acting governor of Kordofan to civil secretary, 1–1–39.
12. Ibid., p. 1295, letter from civil secretary to governor of Kordofan, 11–1–39.
13. Ibid., p. 1294, a note by the financial secretary, 5–1–39.
14. NRO Civ. Sec. 19/1/2, p. 1272, letter from district commissioner of Zande to governor of Equatoria Province.
15. NRO Civ. Sec. 19/1/2, p. 1336 ff., letter from governor of Darfur to civil secretary, 5–11–39.
16. NRO Civ. Sec. 19/1/1, p. 46.
17. Ibid., p. 48.
18. NRO Civ. Sec. 19/1/1, pp. 993 ff., report on famine situation in Dinder area, 28–3–32.
19. Letter from district commissioner of El Geneina to Lieutenant Bertrand, Chef de la Subdivision d'Adre, 25–9–31, NRO Civ. Sec. Darfur, 3/1/1.
20. NRO Kordofan 1/6/25, vol. 1, tenth meeting of the Sudan Emergency Supplies Committee, 18–10–39.
21. NRO Kordofan 1/6/25, vol. 1, p. 52, forty-sixth meeting of the Sudan Emergency Supplies Committee, 28–6–41.
22. NRO Civ. Sec. 19/1/2, p. 1421, letter from governor of Darfur to financial secretary, 17–2–40.
23. PRO J 3469/1711/16, Sudan famine and Egyptian offer of help, 1949.
24. PRO J 6995/1711/16, extracts from reports by governor of Kassala.
25. NRO Civ. Sec. 19/1/2, famine relief cases: relief measures general, p. 1712, letter from Sayed Abdul Rahman El Mahdi to civil secretary, 24–5–49.
26. NRO Civ. Sec. 19/1/2, p. 1298, letter from governor of Upper Nile to financial secretary, 14–2–39.
27. NRO Civ. Sec. 19/1/1, letter from governor of Kordofan to civil secretary, 5–9–32.
28. Interview with Sayed Taha Ahmed, head of marketing section, ABS, Khartoum, November 1985.
29. Interview with managing director of Gash Delta Corporation, 30 October 1985.
30. Interview with director of Crop Market, Gedaref, 7 December 1985.
31. A high level *dura* policy committee was established in the Central Ministry of Commerce to ensure that *dura* production was maximized and that the grain was fairly distributed in the country.

4

Case Studies of Famine: Ethiopia[1]
Hugh and Catherine Goyder

Introduction

The purpose of this chapter is to explain how the Ethiopian famine of
1984–5 came about, and how both the Ethiopian government and
outside agencies responded to it. The first three sections attempt to put
the recent famine in its historical and geographical context. Later
sections discuss the famine itself and examine the conflicts which
quickly arose in the relief effort due to the resettlement campaign.
Finally, there is a discussion of Ethiopia's prospects for avoiding future
famines in the light of recent initiatives like the villagization campaign.

Since the famine the debate about resettlement has become highly
polarized: one has only to compare Cultural Survival's report (Clay and
Holcomb 1985) with the recent book by John Clarke defending
resettlement (Clarke 1986) to understand the two extremes.[2] A second
area of active debate is between supporters of the various liberation
struggles of the Eritreans, Oromos, and Tigrayans and those who are
more critical of these movements. In an attempt to steer a course
through this minefield, this chapter only discusses resettlement, first, as
a serious constraint on the 1984–5 relief effort and, second, as a possible
strategy for preventing future famines. Similarly, the various conflicts in
northern Ethiopia are mentioned, but only in so far as they affected the
famine relief effort and will continue to make the whole area more
vulnerable to future famines. A further omission is the remarkable
initiatives undertaken by the Eritrean Relief Association (ERA) and the
Relief Society of Tigray (REST) in promoting development and
rehabilitating agriculture in areas under their control, which are well
described elsewhere.[3]

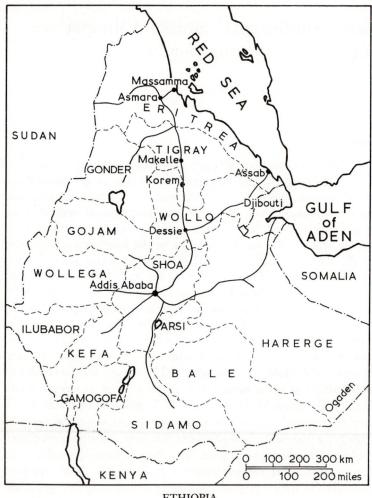

ETHIOPIA

Work on this subject has indicated a number of serious gaps in our understanding about how Ethiopian famines affect individual households and how they cope with famine. Development policy in Ethiopia tends to be based on simplistic assumptions about how peasant and pastoral economies work, and there has also been little research done on the precise impact of the relief effort of 1984–5, though much information must be available in the files of the various relief agencies.

The country

Ethiopia is a vast and complex country. With an area of 1,224,000 sq. km it is roughly the size of South Africa. The country can be divided into three main geographical zones, with the highland plateau divided into terrain above 2,000m (*dega*), a medium highland (*weina-dega*) above 1,500m, and the surrounding lowlands (*kola*). The highlands are bisected by the great East African Rift Valley. The population was estimated at 42 million following the recent census, but it is very likely to be higher in view of the underestimate of population in Tigray and Eritrea and the general unreliability of most Ethiopian statistics. Aseffa considers the total population may be as high as between 50 million and 60 million.[4] Population is concentrated in the highlands, which are free of the malarial mosquito above 2,000m – a factor which partly explains the non-occupation of lowlands suitable for cultivation.[5]

Precipitous terrain and limited road access, together with security problems arising from the civil war in the north,[6] make it hard to penetrate deep into the countryside. It would no longer be possible for Dervla Murphy to make her epic journeys 'with a mule' as she did in 1967.[7] Most of the areas through which she travelled are no longer fully under government control and hostilities have made life hazardous both for Ethiopians and development workers. Government officials are a particular target for the 'liberation fronts' who are active in the north of Gonder and Wollo, Tigray and Eritrea. Life in the south remains more secure and it is here that some progress in rural development has been made since the revolution in 1974.

Ethiopia is a very beautiful country with unique topographical features and a range of wildlife much of which is native to the country. Average rainfall in the highlands varies from 800mm to 2,000mm per year but is unreliable in many areas. The pattern of rainfall is characterized by unpredictability and irregularity. As in other semi-arid areas of Africa, it is not just the total amount of rainfall that is important but its timing and distribution. Over 80 per cent of the population is dependent on agriculture, and the agricultural sector also has to contribute most of the country's export earnings through the production of cash crops, particularly coffee. The majority of agriculturists are subsistence farmers who grow crops or rear livestock. The lowlands are more sparsely populated, mainly by pastoralists who herd livestock, but traditional grazing lands are now under pressure from the government, which has already started irrigation schemes, especially in the Awash valley.

Agricultural life revolves around the two principal rainy seasons: the short *belg* rains from February to May and the long *meher* rains from June to September. The *belg* rains are important in the lowlands for the planting of long-maturing crops such as maize and sorghum. Highlanders use this time to plant short-maturing crops such as barley and wheat. The *belg* rains are particularly important in Wollo, where 50 per cent of production takes place at this time, and in lowland areas like Bale, Sidamo, and Gamogofa. The main *meher* rains are important for the planting of long-maturing crops such as *teff*, a local millet grown only in Ethiopia and of great significance in the traditional diet. The main harvest in the highlands is in November and December.

Ethiopia is one of the poorest countries in the world with a GNP per capita of $US 120 in 1985, which is less than half the average for low-income countries and only just over half the average for poor countries in Sub-Saharan Africa.[8] Health and nutrition standards are poor, although, apart from times of famine, standards of nutrition among children are generally found to be worse in the cities.

In addition to its geographical diversity, Ethiopia is ethnically diverse and it still remains a federation of different national groups. The history of the country is one of a struggle for domination by the Amhara minority, and the same struggle for control over diverse cultures and ethnic groups characterizes the style of government today. Recent conflicts can be best understood in the context of the struggle to build a united people's republic.

The major liberation movements and their approximate areas of operation are as follows:

Eritrea: Eritrean People's Liberation Front (EPLF).
           Eritrean Liberation Front (ELF).
Tigray: Tigrayan People's Liberation Front (TPLF).
Wollo: (Area near Sekota in north-west Wollo) Ethiopian People's
           Democratic Movement (EPDM).

Gonder: Northern and eastern areas are partially under control of EPDM and groups supporting the Ethiopian People's Revolutionary Party (EPRP), which was defeated by Mengistu at the time of the Red Terror (1977).

Wollega: The Oromo Liberation Front (OLF) is active in the border areas.

Harerge: Units of the OLF are active in the remoter highland areas.

In response to these various movements the military government has built up a sophisticated army, heavily dependent on conscription. But it can be argued that the diversity and fragmented nature of the various liberation movements have helped Mengistu consolidate his power in the past twelve years, and they have strengthened the government's belief that Ethiopia's revolutionary socialism is threatened both by external and internal forces.

Agriculture

Ninety-five per cent of Ethiopia's cultivated land area is still worked by individual families. Only 1–2 per cent of farmers are in producer co-ops, though this number is now increasing rapidly. The state farms have never been successful and account for only 4 per cent of the cultivated area. They often face labour shortages at times of harvest and planting, and a number of settlements in the Wollega region were located so that the settlers could work on the state farms when required. Although many outside 'experts' tend to idealize the individual farmer, many of these holdings have become extremely small: as early as 1975, 45 per cent of Tigrayan farmers had *less* than one hectare, and by 1985 this figure had risen to roughly 65 per cent of farmers.[9]

Ethiopia's history of famine

The feudal society

Medieval religious texts refer to famines in the ninth and twelfth centuries; there are further records of four famines in the thirteenth century and other 'cruel famines' in the years 1314–44. In the years 1435–6 a famine is described as having 'destroyed the inhabitants of Abyssinia'. Forty periods of famine have been identified between the sixteenth century and the present day.[10] There are uncertainties about the dates of these famines, but it seems that they were often accompanied by epidemics which killed cattle and oxen and by attacks of locusts or caterpillars which ate the crops.[11] Others were the result of long periods of war such as after the fifteen-year-long campaigns of Ahmad Gran in the 1540s. Furthermore, throughout history, large armies marched their way across the countryside eating their way as they went, for they carried no provisions but relied on the people for food. As Pankhurst has put it, the soldiers were worse than locusts for not only did they destroy what grew in the fields but what had been 'gathered into the house'.[12] There was frequent looting by soldiers and attacks by marauding bandits.

The people were also open to plunder in more direct ways. Until the 1974 revolution Ethiopia was a feudal society. In the south the nobility were supported by the *gult* system, which gave them the right to collect tribute in the form of agricultural produce. This could be anything from one-third to three-quarters of a farmer's crop. Furthermore, the farmer paid one-tenth of his produce as a tax. Not only did the nobility escape payment of these local taxes but they also kept a part of what was collected for themselves. There were yet further obligations for the peasant farmer, who was expected to provide his landlord with honey, meat, firewood, dried grass; he had to grind the landlord's share of the grain, transport it to his residence in town, build his house, maintain his fences, care for his animals and act as porter, escort, or messenger.[13] In the south feudalism was introduced as late as the end of the last century when land was expropriated under Emperor Menelik II. Though the tenure system in the north tended to be different from 'communal' land tenure, that is land belonging to different kinship groups, peasants still had to part with one-tenth of their produce in taxes. They also had to pay tribute, usually one-fifth of their produce, to the agent of the local governor, who was also entitled to ten days' free labour.[14]

As a result of these incessant demands, peasants were kept in a state of semi-starvation. Their obligations to feed others superseded their right to feed themselves.[15] Peasants were left without a surplus to store for times of adversity or to sell in the market-place. Quite apart from the sheer volume of produce being paid as tribute, the amount of time spent working for the landlord resulted in heavy production losses. The nobility remained out of touch with the dynamics of food production and showed no interest in the land until commerical farms were introduced. As a result, productivity remained low and agricultural technique remained primitive.

Life therefore remained extremely precarious for the poor tenant farmer. Even at the time of the revolution the average land holding was only about one hectare and each hectare had to support three people. Tawney's metaphor of a man standing up to his neck in water, so that even a ripple is sufficient to drown him has been aptly used to describe the situation.[16] However, it was the Scottish traveller James Bruce who observed as early as 1790 that rinderpest, locusts, and drought continued to ravage the countryside because the peasants were so poor to begin with. He blamed the socio-political order and claimed that the worst misfortune peasant farmers faced was 'bad government, which speedily destroys all the advantages they reap from nature, climate and situation'.[17]

The pre-revolutionary government not only condemned the peasants to a life of perpetual poverty but also made no attempt to provide for times of famine and there was hardly any relief assistance. Historically, famine had been seen as divine punishment for human misdeeds, and even in 1888 Menelik II reproved his subjects for not praying hard enough. This was the time of the Great Famine, which is reputed to have killed one-third of the population between 1888 and 1892 and 90 per cent of the cattle. The government made no attempt to control the movement of livestock which had been infected with rinderpest. Without oxen, farming collapsed. The situation worsened because 1888–90 were drought years, and caterpillars and locusts destroyed many crops. Food prices rocketed and all those without money starved. It was a time of great social upheaval, and terrifying tales have been told of people being eaten by hyenas, cannibalism, suicide, the sale of children by parents, and self-enslavement. Relief assistance was limited to those who could reach the emperor's palace, where Menelik himself fed the destitute.[18]

Little was done to assist victims of other famines in the twentieth century, who got little relief assistance until public pressure forced the introduction of a relief programme during the Wollo famine in 1973. People in Addis scarcely knew about a devastating famine in Tigray in 1958. Similarly in Wollo in 1966, when the districts of Wag and Lasta were severely affected, the government was not only content to pursue a remarkably leisurely exchange of letters about the situation but also continued to collect taxes when people were already dying. It was assumed that people had money for taxes, and the food which was eventually brought in for sale was allowed to rot and be eaten by rats while the poor starved.[19]

Wollo never really recovered from the 1966–8 famine. Without long-term rehabilitation and rural development, famine seemed to recur wherever there was more than one year of below-average rainfall. In July 1971 the people of Awsa district, or *awraja*, petitioned the governor for food aid and by 1972 nine other districts were facing famine. Tigray, Gonder, and northern Shoa were also affected. According to the records of the Central Statistical Office, the harvest failure of 1973 was attributed to the loss of oxen and the shortage of seeds in addition to the shortfalls in seasonal rainfall. As Wolde Mariam writes:

> Thus the effect of famine on people and livestock, especially oxen, is to reduce drastically the available workforce, which in turn affects the land. It means that all the three factors of production are adversely affected by famine, crippling agricultural production so effectively that it promotes famine on a much wider scale and more intensely.[20]

Most of the Ethiopian highlands suffered from animal and population pressure as people struggled to farm smaller parcels of land. Population increased following the introduction of vaccination programmes and some health and veterinary projects, but there was little uncultivated land into which people would move. Land could no longer be left fallow, and marginal land, often on steep slopes, was cultivated, with the result that soil erosion was accelerated. This increased population required more cooking fuel, and the pressure on forest reserves meant that cow dung, previously used to enrich the soil, had to be used for fuel, thus further impoverishing the soil. A lack of water-conservation measures meant that the heavy rains were left to do great damage – washing away top-soil and loosening stones and boulders, which had been secured by tree roots. These were deposited into fields below, making the land difficult to work and decreasing its fertility.[21]

Although the main group of people affected by the famine were peasant cultivators, there was more starvation among the pastoralists. The failure of the long *meher* rains in 1972 decimated the herds of the Afar nomads who inhabit the Wollo lowlands.[22] The Afar, and poor tenant farmers in other areas, like Arsi south of Addis, suffered from the effects of agricultural 'development' projects aimed at introducing commercial farming. Traditionally, the Afar have depended for dry-season grazing on the Awash River, which floods in August creating an area of rich alluvial soil. Concessions on land previously frequented by the Afar during the long dry season were sold in 1970 to foreign-owned companies and to private individuals, including the feudal overlord of the Afar, Sultan Ali Mirah (who cultivated over 14,000 hectares), for growing commercial crops, cotton, and sugar. The needs of the Afar were totally disregarded and they were forced to use less fertile parts of the valley, which subsequently became overgrazed.[23] It was perhaps the Afar who suffered most from starvation in the Wollo famine of 1973–4 and they were again seriously affected in 1984.

The Ethiopian Ministry of Agriculture's report on crop production for 1972 gave clear indications of impending famine – an illustration of the inadequacy of 'early warning' without early action. Not only were the drought areas accurately identified down to district level but the telling signs of disaster – rising grain prices, an acute shortage of purchasing power, and the falling value of livestock in exchange for grain were reported. The report concluded, 'emergency food distribution may be needed and immediate emphasis should be placed on designing and implementing any necessary distribution programmes'.[24]

But the government did little to act on the situation. A Grain Deficit Study Committee had been set up in November 1971 which eventually recommended in the early part of 1973 that 13,885 tonnes of grain and 4,902 cartons of supplementary food be sent to assist 596,000 drought-stricken people. Requests from regional governors and administrators had been for 46,453 tonnes of grain and 121,340 cartons of supplementary food for a total population of 1,874,047.[25]

The attempts by the government to play down the severity of the situation, and the neglect shown towards the suffering, contributed greatly to the loss of life during the famine. No food reserves were held for relief distribution during times of emergency and the government failed to allocate funds for relief that were commensurate with the scale of the disaster: it was also unable to set up any structure to administer relief. By April 1973 the Grain Deficit Study Committee changed its name unofficially to the National Emergency Committee, but it remained ineffective and did little to co-ordinate the relief effort that finally got under way towards the end of the year. By this time the worst of the famine was over in the North, but Harerge was severely affected. Despite this, most of the food aid went to the north, with Harerge only receiving 8 per cent of food aid.[26]

Thus the 1972–5 famine can be divided into two distinct periods: the Wollo famine from 1972 to 1973 with the peak of the crisis occurring from June to August before the mobilization of a concerted relief effort, and a second phase from 1973 to 1975, when the famine moved south and mainly affected the Ogaden.

Famine has therefore been endemic to Ethiopia throughout history. Nor has the situation improved in recent times. It has been estimated that 'between two and five million' died of famine between 1958 and 1977.[27] Those who died were those at the bottom of the social scale – beggars, prostitutes, beersellers, those who had already migrated from further north, farm labourers, poor relations, and household servants who lost their positions once times became hard.[28] In addition, tenants and small farmers were made destitute by a downward spiral of chronic poverty induced by farming increasingly marginal land and the crippling extortion of tribute and taxation. Families were split as men left in search of work. Above all, the children were vulnerable to the spread of infectious disease, which particularly affected those who reached overcrowded and insanitary relief centres.

For the lowland pastoralist, the 1970s drought brought starvation due to the deteriorating exchange rate of cattle for grain and the drastic reduction in herd size. Among the Afar this situation was particularly due to the loss of traditional grazing land. People often died of starvation, not because of an overall shortage of food but because they lacked the purchasing power to buy food and any legal entitlement to the food which they themselves had produced.

To cope with the overwhelming need for emergency relief, the Relief and Rehabilitation Commission was established in the last months of the *ancien régime* following the resignation of the prime minister, Aklilu Habte Wold, because of public outrage at the concealment of the famine and the appalling lack of relief measures.[29] The revolution followed within six months. Steps were then taken to bring about structural change in the countryside and to introduce measures to ameliorate the problem of famine. That these measures have failed is both a reflection on the enormous inherited problems of a society emerging from centuries of oppression and of underlying weaknesses inherent in the nature of the post-revolutionary government itself. As Wolde Mariam has said:

> To reduce the problem of famine to natural factors or to raise it to an international conspiracy of some sort is to miss the centre of the issue and to exonerate the values and institutions that, both by omission and commission, play a direct role in promoting famine.[30]

We now turn to an examination of the socio-political structure of Ethiopia since the revolution.

The post-revolutionary decade

The revolution has seen the emergence of both positive and negative developments, which are analysed here in an attempt to place the 1984 famine in context.

Land reform and agriculture Land reform remains the most far reaching of these developments and was introduced by the Land Reform Proclamation of March 1975. Privately owned land was nationalized without compensation and regrouped into units of 800 hectares under peasant associations. By 1977, 26 per cent of peasants in Tigray, 23 per cent in Gonder, 41 per cent in Gojam, and 30 per cent in Wollo were grouped into such associations.[31] Farmers in these areas had usufructuary rights to 10 hectares per peasant family, and the communal land system previously practised in the north was abolished. Pastoralists using government land were to have their traditional user rights respected. The reform was, however, implemented by students, many of whom were quite lost away from the capital, and the results were very mixed. On the whole, land reform was more successful in the south, where more land had been expropriated by the crown, more tenants were insecure, and where more mechanized farming was being introduced; it was here that land reform had most support. In the more heavily populated north, where more people had small holdings, there was more opposition to reform, which in itself could do little to solve the problems of over-cultivation and fragmented holdings. Furthermore, land holdings took no account of the quality of land, and access to oxen remains a source of inequality.[32] For instance, it has been estimated that only 36 per cent of farmers had a pair of oxen by 1984 and 30 per cent had no oxen at all.[33]

At least there are 5.5 million farm families now grouped into Peasant Associations and 4.5 million are members of Service Co-operatives which assist groups of Peasant Associations with agricultural inputs, credit facilities, and product marketing.[34] Again, these are better developed in the south than in the north.

Peasant Associations and Service Co-operatives have become important structures for some of the more positive features of rural development. They also provide a framework for political education and military conscription which some observers would count as negative features.

At best Peasant Associations have been able to take on a wide range of initiatives like building village roads and health clinics, and the Ministry of Agriculture relies on them for the implementation of crucial projects in soil and water conservation, most of which involve the distribution of grain donated by the World Food Programme on a Food-for-Work basis. Through the Ministry of Agriculture many NGOs have been able to support a wide range of projects with the Peasant Associations and Service Co-operatives, including the distribution of oxen, seeds, fertilizers, and the protection of springs to ensure safe drinking water.

The criticism that can be made of policies towards agriculture is that, having set up some potentially very useful structures with a potential for rural development, the government is still giving more priority in its allocation both of resources and of technical assistance to Producer Co-operatives rather than to Peasant Associations. This relates to a tendency, by no means confined to Ethiopia, to put more investment into 'surplus' areas at the expense of deficit areas, and too much priority is being given in future development plans for large irrigation projects using rivers like the Awash and the Omo. The disadvantages of these sorts of schemes include their high capital and maintenance cost, their impact on the pastoral people who often lose grazing land, and the fact that the farms created by such irrigation projects will be run by the Ministry of State, which has a poor record of making such state-owned farms economically successful.

Major aid donors like the World Bank also criticized the high level of surplus extraction from small farmers who are not only subject to considerable amounts of taxation but are obliged to sell part of their surplus to the Agricultural Marketing Corporation (AMC), although they sell the remainder on the free market. In order to subsidize grain rations for the urban poor and to feed the army, procurement prices are very low with market prices twice as much, if not more. These restrictions may limit the output of the peasant sector, upon which the government ultimately depends. There are also limitations imposed on the movement of grain between regions.

However, it is possible that the low prices paid by the AMC may not be as significant a discouragement to farmers as has normally been assumed. Only about 20 per cent of the production of the peasant sector enters markets at all, since the AMC does not cover products like livestock or vegetables, and farmers are still able to sell a proportion of their crops on the open market.[35] It is possible that farmers are discouraged less by the price paid by AMC than by the *amount* of their total production which it takes, and that a higher official price would not be the incentive to greater production that some aid donors have assumed. Also, there are doubts about whether in fact farmers would be constrained from increasing production by both the very high price and poor availability of key inputs like fertilizers and consumer goods. Furthermore, there is recent evidence that farmers are now being discouraged from increasing production by a number of factors apart from the AMC's prices: these include the disruptions of the villagization drive, and the increasing amount of time people are expected to spend in meetings.

Economy and war The government has also successfully maintained and developed the country's basic infrastructure – roads, public-sector trucking fleets, telecommunications, and air transport. Furthermore the Relief and Rehabilitation Commission (RRC) has been working since 1974 to set up structures to monitor drought and provide an organizational framework for handling the enormous task of famine relief. Its early-warning system is one of the best in Africa. Without these developments, relief operations would have been greatly hampered in 1984.

On the economic front it can be argued that Ethiopia has done remarkably well with very scarce resources. In many respects it faces similiar economic problems to any other least developed country. The most pressing is a growing balance-of-payments deficit. Export revenues can only come from the export of commodities, mainly coffee, while the very small industrial base means that any investment is severely constrained by the need to import parts and machinery. In spite of this, Ethiopia has followed more responsible macroeconomic policies than most other developing countries. The absence of very large 'prestige' aid projects in the last decade has helped to keep Ethiopia's debts at a relatively modest level, currently at about 26 per cent of GNP, compared with a figure of 52 per cent for other low-income countries in Sub-Saharan Africa. The growth rate of GDP, which was 2.3 per cent per year in the ten years up to 1983, is still less than estimated population growth (about 2.9 per cent) but is also above the average for Sub-Saharan Africa.[36] Between 1974 and 1984 the number of elementary and secondary school children increased from 1 million to 3 million. The number of children of school age attending school increased from 19 per cent to 49 per cent. But it is in the field of adult literacy that the greatest achievements have been made. The National Literacy Campaign claims to have made 11 million people literate and its success was recognized by a UNESCO award in 1980.[37] It is really the demands of the war that have drained resources away from development. There is much uncertainty about the figures, but Ethiopia is spending $440 million each year on defence, which is at least 10 per cent of its GNP;[38] the Soviet Union provides arms and other hardware, but these shipments are viewed by the Soviet Union as being given on a credit basis, and it is thought that Ethiopia's debts to the Soviet Union have reached about $3 billion.[39]

Both Ethiopia's 'dependence' on the Soviet Union and its conduct of its campaigns against the various liberation movements have given it a poor international reputation. But critics forget the grave danger facing Ethiopia at the time of the Somali invasion in 1977, shortly after the United States withdrew both its development and military aid in 1976. Arms from the Soviet Union and up to 10,000 Cuban troops had helped drive the Somalis from the Ethiopian Ogaden in March 1978, and the Soviets were also involved in the bitter battle to regain control of Massamma later in 1978.[40] Many other regimes would have collapsed under these pressures. Having survived, one would expect Mengistu to give the highest priority to the defence of Ethiopia from attacks both from outside and inside. Famine relief and prevention will probably always be a lower priority for the present regime than defence, and even though Ethiopia has now established better relations with Somalia and Sudan, there seems little chance of a negotiated settlement in Tigray or Eritrea.

In these war areas military campaigns are led at times that disrupt agricultural production, and crops and villages are destroyed. Conscription drives remove able-bodied men from the land and may contribute to the decline in food production. Development projects for soil and water conservation and reafforestation have come to a halt in just the areas they are needed most: as a result much of northern Wollo, Gonder, Tigray, and Eritrea are in danger of becoming infertile wastelands. Furthermore, food aid is withheld from insecure areas for fear of them falling into the hands of guerillas and normal government services and institutions are disrupted.

What happens at local level does depend on the interpretation of 'directives' from above. The whole style of government is now tending to focus on a series of campaigns or *zemachas*. Some regions, especially Gonder and Wollega, have had a history of particularly brutal local administration, and regional administrators still have a certain autonomy in the interpretation of directives coming from the centre. Clearly, the priority of the government in September 1984 was that the tenth anniversary of the revolution and the founding of the Communist Workers' Party of Ethiopia be celebrated in great style. Political priorities were paramount and attention and resources became diverted, even within the RRC itself, from the devastating problems of famine building up in the countryside.

Drought and warnings of famine, 1980–4

Just as the 1973 famine grew out of the lean years facing Wollo after 1965, so the recent famine can best be understood as the outcome of several bad years, especially between 1980 and 1984.

In the north, the situation began to decline steadily after 1980. According to Oxfam:

> In 1980 the rains on the eastern escarpment and lowlands of Tigray were 30 per cent of normal, reducing crop yields and killing livestock. The 1981 rains were worse still for parts of the province, though in other parts they were good. According to several reports from Tigray and Eritrea, the failure of the rains meant that, for five or six seasons in succession, the harvest had been poor. ... The central highlands and eastern lowlands of Tigray were particularly hard hit in 1981, and with the failure of the 1982 rains this area broadened into a belt stretching from Adua to Enderta districts. There was scarcely any harvest in some areas of Northern Wollo, Eastern Gonder and central and north-eastern Tigray in 1982.[41]

By December 1982, Save the Children Fund (SCF UK), which had continued to monitor child nutrition in Wollo following its involvement in the last famine, had opened a feeding centre in Korem. This small town on the main road north from Addis Ababa to Asmara marks the 'frontier' between government-controlled and guerilla-held territory, and it began to attract large numbers of people seeking relief food. SCF began with emergency feeding for between 300 and 400 people: by April it was feeding 35,000.[42] The government Relief and Rehabilitation Commission opened grain distribution centres in the northern towns – Makelle, Axum and Adua in Tigray, Ibnat in Gonder, and Korem and Alemata in northern Wollo. However, migrants from Tigray were treated with suspicion and frequently sent away until they could produce the correct papers from their local Peasant Association showing their eligibility to receive relief aid. People who were unable to provide the correct papers found it hard to receive food, and this particularly affected those who inhabited the 'no-man's-land' between government-held areas and those firmly within Tigray and under the control of the TPLF. It was not until the intervention of the International Committee of the Red Cross (ICRC) that people were given assistance regardless of where they had come from.

TPLF attacks on the main road and raids on government property continued to hamper relief operations. In April 1983 the SCF feeding programme came to an abrupt halt following the TPLF raid on Korem. The guerillas ransacked the grain store, shot government officials, and took relief workers hostage. It was five months before the feeding centre was re-opened.[43]

The failure of the *belg* rains in 1983 led to further deterioration of conditions in the north, especially in Gonder where the influx of people from western Wollo in search of food caused intolerable strains on local resources.

Drought conditions led to increasing cattle mortality in the region (as in 1973). It was the Afar in the lowland areas of Wollo who were particularly hard hit. Despite improvements in rainfall in 1983, farmers were unable to carry out normal agricultural activities. Distress migrations began in early 1984 as whole households moved south in search of food.[44] The decision to migrate as a family unit was generally only taken after the sale of all available assets – even, in final desperation, the wooden poles and sticks that had built their homes. Once a family had sold its animals and eaten its seed grain there was very little hope of recovery. It was this destitution that forced many people to accept the only solution offered to them when resettlement campaigns began.

The famine of 1984–5

Over the next year drought became the norm in many parts of the country, including regions in the south, which had rarely experienced drought before. For people in south Shoa, famine was a relatively new experience, and unlike people in Wollo they did not move to feeding centres on main roads. Falling nutrition levels, particularly among the children, was therefore not recognized until it was too late. Welayita Sodo is a densely populated district in northern Sidamo and the situation that developed here in early 1984 became known as the 'green famine'. Since Welayita is traditionally a surplus area, the government continued to levy heavy taxes after the 1983 harvest and farmers were forced to sell food grain, household goods, tools, and animals in order to pay their taxes. When the *belg* rains failed in 1984, as they did throughout much of the country, people were left without any reserves. To make matters worse, even their traditional insurance crop – *enset*, the false banana – was attacked by disease. To the casual observer Welayita after the main rains had begun in July 1984 looked a verdant and fertile area: yet people starved in their own homes and villages, and unlike in Wollo, rarely came into camps on the main roads.

Further south drought conditions affected the nomadic areas of the Hamer and Geleb in Gamogofa and the Borana in southern Sidamo. As in the 1970s, livestock were depleted by lack of pasture and water sources, and it is estimated that more than a million cattle perished in the province of Sidamo alone.[45] Grain prices soared, and the pastoralists became dependent on relief for their survival. Even supplies from the acacia tree dried up, depriving the Hamer and Geleb of the opportunity to collect incense, which in the past had provided them some income with which to purchase grain.[46]

It is now clear that it was the total failure of the *belg* rains which normally come in February and March 1984 that was to transform a series of regional drought problems into a national disaster. As Berhane Gizaw, head of the RRC early-warning section reflected later, 'We allowed them to exhaust all their resources during the bad years of 1982 and 1983 and then left them to face the crisis of 1984 in that state'.[47] Six million people were already estimated to be affected by famine when the final blow came. This was the failure of the *meher* rains, which started late, fell light' if at all, and in all parts of the country were over by the end of August. By then it was estimated that eight million were famine-affected.[48] This famine was far more widespread than the 1973–5 famine, and some observers feel that if conditions had been left to run their natural course, similar scenes to those of the Great Famine a century before would have been the result.[49]

Throughout 1982 and 1983 the government had made repeated requests for international assistance. As early as May 1981 the RRC had presented Ethiopia's case to the UN Conference on Least Developed Countries. By March 1984 the tone of the requests became more grave as RRC Commissioner Dawit appealed to a meeting of all major donors in Addis Ababa for 450,000 tonnes of grain. In fact the RRC had estimated that 900,000 tonnes were required but felt that this was beyond the capacity of the existing transport system to distribute. Why were the RRC's appeals for aid in March 1984 largely ignored? The explanation can be given at two levels, the first at the international level, the second within Ethiopia. At the international level Ethiopia before the famine was not seen as a country of any great importance, and as regards the United States, there were both legal and ideological reasons why Ethiopia could receive no bilateral aid from USAID: legally it was unable to receive official US aid until various cases regarding the confiscation of American assets had been settled, while ideologically both the Soviet involvement in the country and Mengistu's record on human rights confirmed that the Ethiopian government was most unlikely to receive any development assistance from Washington. Although emergency aid from the United States was legally possible, the very limited American presence in Addis Ababa made it difficult to verify the dramatic warnings of famine being pronounced by the RRC from late 1983 onwards: like the other major donors, the Americans had to place excessive reliance on the United Nations agencies in Ethiopia, which themselves had very limited access to the famine areas.

As regards the EEC, the response to the famine was muted more by bureaucracy than any strong ideological factors, although Ethiopia had plenty of enemies from different political parties in the European parliament. The real problems were first the immense delays between a decision to allocate grain and actually delivering it and, second, the fact that most EEC food aid is 'locked in' to agricultural development projects in which it is used as Food-for-Work. This sort of food aid was always added to estimates of stocks of food aid available in Ethiopia, but in practice it was often under the control not of the RRC but of the Ministry of Agriculture, which stored it for its own projects far from the key famine areas. It was only at the personal request of Dawit, the Commissioner for the RRC, during his visit to Brussels in May 1984 that 18,000 tonnes of EEC food aid were reallocated from Food-for-Work projects and released for famine relief.[50]

Yet some parts of the Ethiopian government must accept some responsibility for not giving more support to the RRC's request for massive aid to combat the famine in 1984. In what is loosely referred to as the 'international donor community' in Addis Ababa there was little understanding about the realities of life in the countryside in general, and the inaccessible and war-affected north in particular. Many UN and diplomatic staff tended to rely on the NGOs for their information, but development agencies, especially the church organizations, had a much better understanding of conditions in the south of the country and very few of them had major programmes in the worst affected areas of the north before 1984. To support the RRC's appeal in March 1984, the Ethiopian government needed to encourage people to visit Wollo and study the conditions for themselves; instead, such travel was severely restricted in July and August 1984 while the government was preparing to celebrate the tenth anniversary of the revolution in September, and during these months considerations of national security were predominant.

In the absence of first-hand evidence of the drama that was beginning to unfold in northern Ethiopia, the World Food Programme (WFP) and the RRC became locked in an increasingly irrelevant argument about grain statistics, which was not helped by the report of the FAO/WFP mission of March 1984, which was finally published in June 1984. The FAO/WFP mission comprised a team of 'international experts', who spent three weeks visiting accessible parts of Tigray, Eritrea, and Gonder. No field visits were made in Wollo or in the south.

The mission accepted the gravity of the situation facing the country and agreed with the government's estimated needs for 605,000 tonnes of relief food, but the report concluded:

Considering the present and potential logistical capacity for the transport and distribution of relief goods [mainly food aid] as well as the accessibility to the affected areas, the mission concluded that 125,000 tonnes can be transported from the ports to the drought affected areas and distributed internally between April and December 1984.[51]

The handling capacity of the port of Assab was always a bone of contention between the RRC and its food donors, and it sometimes seemed as if the prospect of grain being stuck at Assab was more of a nightmare for some UN officials than the thousands of people starving inland. In fact, the government was to prove wrong the United Nations' estimates that only about 1,200 tonnes of grain could be handled by Assab in a day: by the end of 1984 Assab was able to handle 3,000 tonnes a day, and about 1 million tonnes in the whole of 1985. Thus the report was right to mention Assab as a potential bottleneck in Ethiopia's ability to receive food aid, but entirely wrong to base its recommendations for future food aid requirements on theoretical constraints, which were never to prove as serious as the WFP had feared.

The mission's report was not only wrong in itself, but the confusions it generated were even more damaging. As the RRC did not accept the mission's estimates, the publication of the report was delayed three months while the debate continued between February and June 1984, and during these months many UN staff were still asserting privately that there were still adequate stocks of grain with the Agricultural Marketing Corporation (AMC).

There was a further irony in one of the FAO/WFP mission's recommendations for the 'strengthening of the national Early Warning System including the completion of the agrometeorological station and communication networks'.[52]

In retrospect we know that the system was operating efficiently and as a result the government was requesting very large amounts of emergency food aid. As Taye Gurmin, the deputy commissioner of the RRC, explained somewhat despairingly, the only objective of the early-warning system was '*action* on behalf of those affected by food shortages. If action does not result, there is little point in the system itself'.[53] Not only was the international community unprepared to take the government seriously but it would not allow the RRC to stockpile emergency food reserves which could have helped to save lives.

By June 1984 the enormity of the crisis that was emerging began to dawn on the voluntary agencies based in the country, but there was still little response from the international donor community. The government then became obsessed with the importance of its tenth anniversary celebrations. By August the RRC announced that it had run out of grain. No stock had been available since July and pledges of roughly 87,000 tonnes of grain and 8,000 tonnes of supplementary food had still not arrived.[54] By this time the voluntary agencies had begun to realize the extent of the disaster. In August 1984 Oxfam decided to organize its own grain shipment, largely as a result of its loss of confidence in the ability of the main food donors to meet Ethiopia's needs. The cost of this shipment was far beyond Oxfam's own resources, but a number of other agencies, most notably the Norwegian Save the Children Fund (*Redd Barna*) and Norwegian Church Aid contributed, and in September 1984 the British government agreed to add 3,300 tonnes of British grain to the 10,000 tonnes which were being purchased and shipped by the agencies.

In September 1984 the Christian Relief and Development Association, on behalf of twenty-six churches and voluntary agencies, telexed an appeal to the UN and donor governments for immediate relief assistance. As the RRC appealed yet again for assistance in October, the voluntary agencies turned to the media for support in publicizing the famine. As a result the BBC gave prominent coverage on evening news to the newsreel made by Michael Buerk and Visnews photographer Mohammed Amin. The quality of the film was of such intensity that it was subsequently shown by 425 of the world's television networks. At last famine pictures were shown in the United States, where hunger in Africa had until now rarely been of much media interest. The opening words of Buerk's commentary helped to illustrate the dimensions of what was taking place: 'Dawn, and as the sun breaks through the piercing chill of night on the plain outside Korem, it lights up a biblical famine, now in the twentieth century. This place, say workers here, is the closest thing to hell on earth'.[55]

In all the public interest that following the wide distribution of the BBC film of the Ethiopian famine, it was easy to forget that the media had stumbled over this disaster almost by accident. What would have happened if Michael Buerk, or the BBC's East African correspondent, Mike Wooldridge, who was also involved, had been diverted to another story; or if, having covered the story, it had been squeezed out of the evening TV news by another international or even a domestic news story? Undoubtedly, the major relief effort of 1984–5 would have been further delayed and neither the airlifts nor Band Aid would have happened.

The involvement of the media in the Ethiopian famine emphasizes that 'early warning' of droughts is all about sounding effective alarm bells which ring so loud that they secure a response: in the Ethiopian case TV secured a response where the RRC's figures had failed. Both the films and the public response were so powerful that the problem now is that more recent disasters, like the Mozambique civil war and famine of 1987, are, in strictly media terms, an anti-climax.

Relief and resettlement, 1984–6

In the last three months of 1984 a massive relief operation finally got under way in all the famine-affected areas which the government could reach. In view of the massive international resources put into this operation, it is surprising how little is known about the precise impact the relief effort had. For most of the aid agencies it was difficult enough to ensure a regular supply of food to their relief centres and little research was done on the impact of the relief aid given to the people fortunate enough to receive it.

By mid-1984 the famine was most acute in the northern regions, especially those areas already suffering civil war: Eritrea was always able to receive a higher proportion of its total food requirements than other regions because it has its own port, at Massamma, and the government was keen to show it could meet the food needs of the few towns under its control.

In the areas held by the liberation movements, the EPLF which controls the north-eastern part of Eritrea, suffered a grave shortage of trucks but was later able to develop a supply line from Port Sudan. The TPLF and their relief agency, the Relief Society of Tigray (REST), had much more problematic supply lines through eastern Sudan and reported in March 1984 that they were only able to feed 6 or 7 per cent of the 1.5 million people who needed immediate relief. As a result, there was a massive exodus of Tigrayans from Ethiopia into refugee camps on the Sudan border in late 1984. Within Tigray the Ethiopian government only controls the capital, Makelle, and a few other towns like Adua and Axum. These places can only be supplied by air, or by irregular convoys travelling with the Ethiopian army. Thus the reason for the airlifts organized from November 1984 onwards by the British, Soviet, and other air forces was not the lack of roads, as there is an adequate Italian-built road all the way from Addis Ababa to Asmara via Makelle, but the civil war.

In late 1984 there was still far too little food reaching some of the major famine areas of Wollo. Initially, agencies thought this was caused by the very real logistical problems at Assab, the Red Sea port which has to handle the bulk of Ethiopia's imported goods. The failure of the main food donors to respond to the RRC's appeal of March 1984 meant that stocks of grain in the country reached critically low levels between June and October. Once food started arriving in larger quantities in the last few months of the year, the relief effort ran into immediate bottlenecks, not just the limited capacity of Assab to handle bulk-cargoes but also a shortage of transport to take the grain to the main areas of need. A third serious bottleneck was the lack of short-haul, four-wheel-drive trucks to take the grain to the many places off the main roads. Donors tended to underestimate the time required to mobilize a trucking fleet in a country like Ethiopia; it was to take up to twelve months before some of the trucks ordered by the UN and other agencies were finally on the road.

The relief operation had major problems of its own, apart from those caused by the resettlement programme. People's needs in a famine are remarkably simple: they just require a basic grain ration of at least 500 gm per person per day, and in Ethiopia the food aid wheat from the EEC or other sources was perfectly suitable to meet the needs of the majority of the population, but there was rarely enough to go round. Children in Ethiopia are brought up on a largely cereal diet, supplemented by small quantities of cow's or goat's milk and, on special occasions, meat. It is only when the basic grain ration is insufficient for the family that the children's nutritional status suffers, and can usually only be revived with supplementary food of a higher calorific value like skimmed-milk powder, sugar, and oil. Yet the airlifts from both western and eastern bloc countries that started arriving at the end of October 1984 carried all manner of relief materials – milk powder, clothes, blankets, toys, medicines and special biscuits – but they did little to help solve the major cause of starvation, the lack of a basic grain ration for up to 8 million people.

It had been TV pictures of Korem and Makelle that had captured the world's imagination, and a very large part of the relief effort was to go into helping people in these camps which appeared at many points on the main road from Addis Ababa to the north. But these camps were only a small part of the reality of famine: in the areas further south, where both communications and security were much better than the north, people went hungry but were unable to receive relief rations near their own homes. The famine alone was disaster enough, but the congregation of starving and frequently sick people in camps with poor water supply and inadequate rations was a secondary cause of much mortality.

In the response of the relief agencies to this crisis, two approaches emerged. The majority of smaller agencies, while aware of the priority that had to be given to general food distribution, concentrated on medical aid and supplementary feeding in the relief camps, and assumed that the RRC would look after the general distribution of grain. For instance, the 13,300 tonnes of grain imported by Oxfam and the Norwegian agencies had on arrival been given to the RRC for distribution. The American agencies preferred to set up their own distribution systems, by which they took full responsibility for shipping, transporting, and distributing grain, usually in the form of family rations programmes. But all the agencies faced a common difficulty in that almost all their grain had to pass through the congested port of Assab.

The agencies which relied on the RRC to supply a basic ration while they concentrated on supplementary feeding and medical relief work soon found themselves in a very difficult position. In many places off the main roads supplementary food was being given out, but there was nothing to supplement. Frequently, nutritionists use a 'cut-off' point in selecting children for feeding, so that only children who are 80 per cent or less of their normal weight are given supplementary food. But with irregular general rations distributions from the RRC, children discharged from the supplementary rations programmes soon lost weight; and the programmes tended to become pointless. On the other hand, the family rations programmes worked better in the short term, but raised questions about whether relief agencies should set up their own distribution systems or whether they should choose the often more frustrating course of trying to make existing government structures more efficient. The RRC has argued that it has been weakened by the investment of the major American agencies in their own transport and food distribution systems; but the counter-argument is that the RRC risked being overwhelmed by the belated response of the international community to all its appeals, and in the short term it could not expect to remain a 'monopoly supplier' of relief goods. In the long-term, future famine prevention depends on the RRC retaining some ability to co-ordinate relief distributions.

By early 1985 both the UN and voluntary agencies became concerned that the needs of the main famine areas in Wollo were being sacrificed to enable the RRC to send grain to the new settlement areas in the west.

Resettlement

Given the vast amount of criticism the resettlement programme has faced, it is necessary to spell out both the theory behind it and the conditions under which it might assist Ethiopia's food production. Historically, both Ethiopia's people and its rulers have been relatively mobile: Aseffa mentions how, in the sixteenth century, the Oromo moved into the central and northern highlands.[56] In times of famine it is normal for people to leave their homes in the northern highlands, seeking either work in the sorghum-growing plains of eastern Sudan or relief food in the towns.

For a settlement programme to work it would need to be *part* of a development strategy that gives high priority to conserving the environment, and it would have to be combined with major efforts to conserve soil, water, and forest resources both in the areas from which people are coming and in the areas to which people are being moved. Second, it would have to be a very long-term process, as much research and planning has to go into identifying settlement sites and equipping them with basic infrastructure. It is also an extremely expensive exercise and ideally the costs should be spread over five to ten years. Third, a successful settlement programme requires the consent both of those being moved and of the inhabitants of the settlement areas. Fourth, and related to all the previous conditions, the programme should ideally have some kind of international funding and technical support to enable it to be of a sufficient scale to have some kind of environmental impact.[57]

Few of these preconditions of success applied in Ethiopia. In September 1984 the government decided to move 800,000 families, or about 3 million people, from northern Ethiopia to supposedly more fertile areas in the south and west. Some 300,000 families, or about 1.5 million people, were to be moved in the next year. The resettlement campaign was to be implemented by the local officials of the Workers' Party of Ethiopia, and there was to be much competition between Party officials and different administrators of the different sub-districts, or *weredas*, to see which of them could quickly meet the quota of resettlers for their particular area. Although resettlement has a long history in Ethiopia, the new campaign marked a break with the past in that this time the head of state, Mengistu, was to throw his full weight and that of the newly formed Workers' Party behind the campaign. The RRC, which had been officially responsible for resettlement since 1979, was no longer to control the new campaign. It was admitted that 'mistakes' had been made in the resettlement programme in the past, and donor agencies were assured that the new campaign would be entirely 'voluntary' and that settlers would receive individual holdings in the new sites and would not be forced to join collectives or state farms.

Thus the government saw the relief effort as the immediate solution and the resettlement campaign as the long-term answer; no conflict was foreseen between the two activities.

However, the two programmes were to come into conflict at many points, and events later in 1985, including the defection of Relief Commissioner Dawit Wolde Giorgis and the expulsion of the French agency Médecins sans Frontières (MSF), can be best understood as the direct results of this conflict.

Undoubtedly, the first stages of the 1984–5 resettlement campaign involved some loss of life, as people were taken out of relief camps in Wollo and Tigray in a very weak condition and an unknown number died on the long journey south. Senior officials of the Party in the resettlement areas had not been consulted about the food situation in their areas at the time the resettlement campaign was announced. In fact, the famine had hit these areas as well, and the people living in these areas had very little food to share with the new settlers. But once the Politburo had committed itself to resettlement, the settlers had to be fed, and limited resources both of trucks and of food had to be diverted to the settlement areas.

Conflict between relief and resettlement was most evident in Wollo and Tigray, where relief food was used as an 'incentive' to get people to resettle. In most of the sub-districts people who wished to be resettled or those who could be persuaded to join the campaign were first brought to transit camps, which were usually established near the administrative offices. When food arrived, priority in distribution was given to people waiting in these transit camps until transport could be found to take them to the settlement areas. Food was used as an inducement to persuade people to commit themselves to resettlement, and the Party used the logistical problems of the relief efforts to its advantage. The Party's line was that relief supplies would always be inadequate and that if people volunteered for resettlement, the government would 'look after' them. In Tigray there were a number of stories of people being enticed into government-controlled towns by rumours of impending grain distributions, and then being taken away for resettlement, and there is some evidence that the targets in these operations were young men who were felt to be the most likely group to join the TPLF.

The relief agencies working in Wollo and Tigray were not silent in the face of examples of forced settlement. Most of the agencies protested about the abuses of human rights which they encountered, but it was not always clear at what level of government protests should be directed. For instance, a common problem was the use of relief agency trucks to transport settlers from outlying areas to the main transit camps from which people would then be taken direct to the settlement areas. Government trucks were supposed to be used for this purpose, and there were always strong protests when trucks owned by relief agencies were forced to carry settlers. The agencies concerned took up these issues very frequently both at regional level and at meetings with government ministers in Addis Ababa. They also worked closely with the UN Emergency Office, which by early 1985 was providing a reasonably effective co-ordinating role under the leadership of Kurt Jansson. On several occasions Jansson was able to express the agencies' concerns to Mengistu and other members of the Politburo, but we will never know how much the slowing down of the resettlement campaign at the end of 1985 can be ascribed to his representations.

There is some evidence that the degree of compulsion used in the resettlement campaign did increase in the last three months of 1985. There was a respite during the main rainy season, which disrupts communications from July until late September, but once the rains finished, resettlement was resumed with unprecedented vigour. The continuing high profile being given to resettlement in mid-1985 was evident in the Politburo's decision to close Addis Ababa University several weeks early to enable all the students and staff to spend an extended long vacation helping construct new settlement areas. But the main rains of 1985 had drastically altered the situation in the areas from which the settlers were being drawn; for the first time in three years farmers were optimistic about getting a reasonable harvest. In addition, by late 1985 the relief agencies' own transport capacity was much improved, and more grain and other relief food was reaching Wollo. Local officials must have realized that the resettlement campaign was not going to meet the ambitious targets laid down by the Politburo, and in response they made the campaign more 'compulsory'.

The great irony of the final months of the resettlement drive was that the pressure to meet quotas was so strong that more people were taken from the better-off districts of central Wollo, easily accessible from the regional capital of Dessie; less people were taken from the most overpopulated and historically drought-affected districts of Wag and Lasta in the north-western part of Wollo, as the government had less control over these areas, and the only easy source of settlers lay in the relief camps like Korem. Figures produced by the United Nations in 1985/6 compared the numbers resettled from each district in Wollo with the numbers officially described as 'affected' by the drought[58]. These figures show that 21 per cent of those affected by drought in Dessie Zuria, close to the regional capital, Dessie, were resettled. In contrast, only 7 per cent of those affected by the drought were moved from Wag and Lasta, the areas probably most acutely affected by the famine.

At the end of 1985 resettlement suddenly stopped. The reasons have never been entirely clear: the government's position was that it had completed the first stage of the campaign and it now wished to consolidate the new settlement sites before sending in more people. The international community preferred to believe that the many representations made to members of the Politburo through diplomatic channels and the UN Emergency Office about the abuses of the resettlement campaign had led to its suspension after fifteen dramatic months. The French agency MSF, which had protested so publicly about the campaign that it had been expelled from Ethiopia, might also feel that its protests had played some part in the moratorium on resettlement. However, there was little self-congratulation among the voluntary agencies left in Ethiopia, as the suspension of the campaign came too late to prevent further weakening of international confidence in Mengistu's regime and resettlement was expected to resume later in 1986. It was significant that several of the senior officials in the government whose work brought them into the most contact with the international community defected in late 1985 and early 1986. Dawit Wolde Giorgis, the able commissioner of the RRC who had always been considered close to Mengistu, left for the west in October 1984 and never returned. His deputy, Berhanu Deressa, took a strong line with MSF but himself defected after a few months. Then later in 1986 the foreign minister, Goshu Wolde, also defected, and in December 1985 Cultural Survival produced its damning report on resettlement based on interviews with Ethopian refugees in Sudan.[59]

Was resettlement ever a viable and defensible policy for Ethiopia? Most of the criticisms made by agencies like Cultural Survival and MSF related to the human rights abuses that resulted from the campaign, and these have now been well documented. But these criticisms refer much more to the aggressive style with which the Party sought out settlers. They do not attack the idea of resettlement as a valid *option* which could have been pursued in a far more sensitive and gradual manner.

In the period 1974–84 about 200,000 people had been moved into settlement sites. While most had come from the northern regions, a minority had come from Addis Ababa and other cities. The 'conventional settlements', in which there was a mainly collective system and a heavy dependence on tractors, had largely been failures, but this was by no means the only model that had been tried. An alternative was the 'low cost' model which had worked well at Tadelle in western Shoa in which families were simply given a plot of land, hand tools, and a grain ration for one year. A third option, which the RRC developed in 1983, involved settling people in existing farmers' associations in parts of the south and west of the country. This third model was to be much used in the resettlement campaign of 1984–5.

However, the focus of eternal criticism has always been on the large 'conventional settlements' like Asosa (in the province of Wollega), which have never been successful in spite of heavy investment both by the RRC and by outside agencies. The reasons for failure relate to the over-reliance on imported machinery, the collective system of farming, and unimaginative and ecologically disastrous cropping practices. For instance, Asosa relies almost entirely on a single crop, maize, which is highly vulnerable to pest attacks. Life in the larger settlements tends to be very much more regimented than in the smaller settlements built within existing farmers' associations. There is evidence that the major problem settlers faced in these more 'informal' settlements in 1985 was simply neglect. The local Oromo population was expected to build houses for the new arrivals, and to share their very limited food with the settlers. There were rarely sufficient funds to enable the settlers to be given oxen, and in the first few months they were very dependent on the sub-district administration. Here those Amharic speakers from Wollo who spoke the same language as the local officials had a great advantage over the Tigrinya speakers from further north, who usually speak neither Amharic nor Orominya. The Cultural Survival report confirms that it was the Tigrayans who were most determined about escaping from the settlements.

The financial costs of resettlement are really unknown, as these were spread around the many different departments of the government, like transport, health, and agriculture, which had to participate in the programme, and very large contributions were made by the people living in the settlement areas. But there were further indirect costs which are still being paid in the form of foregone foreign aid. The excesses of the resettlement programme made donors reluctant to make long-term commitment and the villagization programme (outlined below) was to discourage even Ethiopia's traditional supporters, like the Swedish government. One must add to the balance sheet the immensely high human costs of the programme – with families split up and even in some cases forced to leave the fields they have always cultivated.

On the other side of the balance sheet it is worth recording that resettlement has helped a large number of people to start a new life. Few surveys have been done on this subject, but anecdotal evidence suggests that there was always a minority of people, especially at the height of the famine in late 1984, who were only happy to be moved, though undoubtedly, as with Irish emigrants to America in the nineteenth century, their enthusiasm was to an extent based on an over-optimistic vision of life in the settlement areas. There was also a significant group of people who moved rather reluctantly and probably under some pressure from their Peasant Associations and Party officials, but who are now determined to make a better life in the settlement areas. A third group of people, including all of those interviewed by the team from Cultural Survival, did not want to be moved and have since either become refugees in Sudan or returned to their original areas. One would expect this third group to be more significant in the more regimented settlements like Asosa referred to above; while in the 'low cost', smaller settlements the second group, the reluctant but hopeful settlers, are likely to predominate.

Firmer evidence about the resettlement programme has become available in an evaluation commissioned by the Irish agency Concern about its programme in Jarso and Ketto settlements in Wollega.[60] The evaluation team included one official of the RRC, and the evaluation was done with the active co-operation of many RRC and Party officials; but the report is highly critical of the ecological impact of resettlement. The rapid implementation of the programme, and a tendency to measure progress in terms of the area of land cleared, has led to rapid reduction of the forest resources in the immediate area of the settlements; and the evaluation team is deeply worried that the current approach involving rapid land clearance followed by an emphasis on growing cereals, will have the same serious consequences on the environment that agricultural systems have had in Wollo and Tigray. Under this analysis, resettlement may not be a solution to Ethiopia's famine problems in the long term, but may just help spread the related environmental problems from the north to the south and west of the country.

These long-term environmental costs may well be concealed by the fact that in their first two years most of the settlements have not been disasters in terms of grain production; yet considerable uncertainties remain on vital issues, such as whether livestock will be owned individually or collectively, and how future crop surpluses will be distributed. The new settlements have been greatly assisted by the fact that average rainfall figures in 1986 were well above average, but it is hard to avoid the conclusion that for all the limitations of their lives a number of settlers in places like Jarso are rather better off than if they had stayed in some of those areas of Wollo and Tigray acutely affected by the 1984 famine. It becomes rather more difficult to work out the impact of the resettlement campaign on the areas from which people have come.

Table 4.1 shows the number of people resettled in relation to the rural populations of Wollo, Tigray, and Shoa.

Table 4.1 Numbers and proportion of rural populations resettled

	Total rural population (millions)	Numbers resettled 1984–5	% of rural population resettled
Shoa	7.3	112,442	1.5
Wollo	3.4	386,000	11.3
Tigray	2.2	89,343	4.0
Total	12.9	587,785	4.5

Sources: J. Clarke, *Resettlement and Rehabilitation*, 1986; RRC, *The Challenges of Drought*, 1985.

It is first important to note that according to the RRC's figures more people were taken for resettlement from Shoa, which was the area best covered by relief agencies, than from Tigray. This supports the hypothesis that the resettlement quotas could only be met by getting people from the more accessible rather than the most drought-affected areas. But the real argument must be over whether the removal of 4 per cent of the highland population will have a positive or negative effect on food production: at present we have no evidence on this. In any case, in rain-fed agriculture the amounts and timing of rainfall are likely to have a far greater impact on agricultural production than marginal decreases in population. Also, if one assumes a birth rate of 3 per cent a year, it will take only four years to replace those removed by the resettlement campaign, even in Wollo, where 11 per cent of the population has been removed.

A further defence of resettlement has been the lack of alternative solutions to the problems of famine in Ethiopia. The complexity of Ethiopia's geography, already described, does make it inherently unlikely that any *one* strategy will make future famines less likely. The lowland pastoral areas will require a very different famine-prevention strategy from that for the settled highlands. Within Wollo, for instance, the only sensible solution will be to follow a wide variety of development measures all aimed at conserving the scarce resources of soil, water, vegetation, and forests. These varied measures, including fencing-off hillsides, tree-planting, terracing, and growing more fodder crops, all require a high level of popular participation if they are to have a chance of success. This in turn raises issues of finance, income support, and grain reserves. The campaign (*zameche*) approach adopted by the Party as a means of achieving its objectives is unlikely to work unless much more effort is made to get enthusiastic co-operation of the people concerned rather than the present grudging acquiescence based on fear.

As regards incentives to agricultural production in Ethiopia, until recently aid donors like the World Bank have stressed mostly the key role that should be played by better official prices for cereals and liberalization of trade to allow grain to move more freely from one region to another. But this liberalization is unlikely on its own to help the people subsisting in drought-prone areas like Wollo, who would need a 'package' of measures to help them increase their production. Probably their first requirement would be assistance over a number of years with agricultural inputs – seeds, oxen, and fertilizers, all of which could be distributed through the existing structures of Peasant Associations and Service Co-operatives. Next, an extension service would have to be designed with the sole objective of helping farmers diversify and increase their production: through this service it should be possible greatly to expand the range of crops grown in the highlands during the *belg* and *meher* rains – at present, potatoes, sweet potatoes, and other vegetables all look promising. Third, there would need to be a rapid improvement in basic water supplies, both for drinking and for very small irrigation schemes, as there is no chance of a more intensive agricultural system having any impact while women in particular have to spend most of the day carrying water.

Villagization

The recent policy of villagization seems highly unlikely to reduce Ethiopia's vulnerability to famine, and the 'campaign' of 1985–6 can be seen as extension of the 'coercive' approach. Significantly, the policy was initially introduced in the province of Bale to help the government regain control of the area after the Somali invasion. It was then taken up by Party officials in Harerge in April 1985 in response to the activities of the Oromo Liberation Front. The idea is to bring together the traditionally widely scattered settlements of the Ethiopian highlands into large, organized 'villages' which can be provided with basic services like schools and health clinics. In practice the move makes government control very much easier. It will be harder for farmers to conceal any grain surpluses they have from the AMC. The Politburo judged the villagization programme in Harerge a success, and from mid-1985 onwards the idea was implemented rapidly in the major grain producing regions of Shoa, Arsi, and Gojam, and new villages have now been built in most of the country except for Wollo, Tigray, and Eritrea. Mengistu announced in his Independence Day speech that in September 1986 over 4 million people, or 10 per cent of the population, were living in the new villages.

Villagization may well make collectivization of agriculture easier in future, but in most places land is still cultivated individually. However, the long-term impacts of villagization on food production must be negative. First, the programme has added to rural impoverishment. People have had to give up their main asset, their house, and the small gardens that surround these *tukuls*, especially in the areas of higher rainfall. Second, crops are now more vulnerable to attacks from birds, monkeys, hyenas, and other pests as people usually have to return to the new villages at night. More time is spent in walking to and from the fields. Third, villagization has had an immensely high environmental cost as much vegetation has been sacrificed both in the building of the new houses and in the clearing of new sites. Fourth, as with resettlement, villagization has added to the scepticism with which the rural people regard the government. There has been surprisingly little active resistance to the campaign, but in the long term a more cynical rural population is unlikely to listen to Party exhortations to grow more food and cash crops. The very long time lag that will now elapse between the move to the new villages and the arrival of the promised 'services' like schools, health clinics, and water supplies will add to this scepticism.

Conclusion

At the time of writing all the potential dangers of the villagization drive have been concealed by the excellent rains of the 1986 crop season, which resulted in some of the best harvests for several years; and a minority of farmers may have welcomed villagization in any case. But the small resources of the Ministry of Agriculture are still being directed to settlement areas, state farms, producer co-operatives and to areas of high agricultural potential rather than to the famine-prone regions. With rapid population growth the per capita availability of cereals has fallen by about 22 per cent in the last ten years. Future famines are probable if there are a series of years as bad as 1983–4. Cultural Survival have gone so far as to argue that both the 1984 famine, and any future famines, are largely the government's responsibility. They accuse the government of 'Establishing a social and economic system that will produce starving people for generations to come'. They have further attacked the role of all humanitarian agencies in Ethiopia on the grounds that 'humanitarian assistance, with no questions asked, helps the Ethiopian government get away with murder'.[61]

This view that the government was responsible for the 1984–5 famine cannot be sustained in the light of Ethiopia's long history of famine described in this chapter. It also dangerously underestimates the critical part played by environmental factors in this history. The resettlement campaign of 1984–5, which is really the target of Cultural Survival's study, cannot be represented as a 'cause' of the famine since it was only started as part of the government's response to it. One can agree with Cultural Survival's criticisms of the resettlement campaign, and deplore the abuses of human rights which have been documented, but it seems highly irrational to deny relief and development aid to a nation of 42 million people because of the government's policies towards the 1.5 per cent of the population, or 600,000 people, who were resettled. There is further scope for argument about whether the withdrawal of western aid and foreign personnel will lead to a softening of the government's stand or have precisely the opposite effect.

What both the supporters and opponents of the Mengistu regime may be neglecting is the surprising continuity in Ethiopia's history, both in its uncompromising agricultural policies and in its policy towards liberation movements. What links the current government in Addis Ababa with its imperial predecessors is a view of the peasantry which sees them as a small and relatively unimportant part of a grand design. Both Haile Selassie's Ethiopia and Mengistu's Socialist Republic have been unable to devise agricultural policies that will either motivate individual farmers or provide food security. For the rural majority, the tax demands of the current government have simply replaced the rents of their previous landowners. The replacement of a feudal system by a revolutionary one has not yet allowed the development of real famine-prevention strategies which would have to take account of the great ethnic and geographical diversity of Ethiopia.

The paradox faced by Ethiopia today is that the more extreme or Stalinist its rural policies become, the more it will have to depend on western food aid. With the RRC's excellent early-warning system, well established and tried, and the large numbers of aid agencies working in Ethiopia, it is unlikely that future famines will go unnoticed. Depending on rainfall, the country will have a long-term deficit in cereals of between 500,000 and 1 million tonnes a year. This dependence on western aid may well lead to splits in the Politburo between pragmatists keen to reduce this dependence and the hardliners.

For the future it seems possible for the international community to help prevent droughts turning into famines without necessarily supporting the more extreme policies of the Ethiopian government like villagization. A simple policy measure that would make an immediate difference would be the establishment of a network of regional grain reserves under joint RRC and United Nations control. The idea of strategic grain reserves was being pursued by the RRC long before the 1984 famine, but the food donors were always reluctant to supply sufficient grain to fill the reserves. The weak point in the RRC's early-warning system has always been not the collection of data but the difficulty of responding to the early warnings with enough distributions of relief grain to stop the situation getting worse. Following the famine, the country's transport capacity has been vastly increased, and provided both the government's and the NGOs' transport fleets can be maintained at about their present levels, there is no reason why food cannot be quickly transported to areas in need; the constraint in the future may well be not lack of either transport or grain but the increased controls imposed by the Party on the RRC, which could make it more difficult to take the quick decisions that are vital in drought relief work.

At an international level the Ethiopian famine caused a painful, but rapid, learning process in both the United Nations and voluntary agencies. The establishment in 1985 of the UN Emergency Office for Africa with an office in Addis Ababa was successful as a co-ordinating body in bringing together, albeit temporarily, the different and often conflicting agencies of the United Nations. Voluntary agencies became suddenly aware both of their power and of their responsibilities in bringing famine to the attention of the world; many agencies now have their own 'in-house' early-warning systems, and the majority of the agencies which took up relief activities in Ethiopia stayed on to take up development work once the famine was over.

On a more pessimistic note, one must emphasize that the northern provinces most at risk to future famines are still deeply affected by civil war, and there seems little hope of these wars ceasing before the end of the century unless the two superpowers, which until now have been fuelling the conflicts, agree together to persuade the parties in the conflict to begin negotiations. Recent improvements in the relationship between the United States and the Soviet Union indicate that it might soon be possible for the two countries to start talking about the long-forgotten conflicts like those in Tigray and Eritrea. Without some negotiated settlement to these wars the Ethiopian famine will seem to most of the people in these provinces no more than 'a period of greater misery in a prolonged age of suffering'.[62]

Notes and references

1. The views expressed are those of the authors and not of their employer, Oxfam.
2. Jason W. Clay and Bonnie K. Holcomb, *Politics and the Ethiopian Famine 1984–85*, Cultural Survival, December 1985; John Clarke, *Resettlement and Rehabilitation: Ethiopia's Campaign Against Famine*, Harney & Jones, 1986.
3. For Tigray, see M. Peberdy, *Tigray: Ethiopia's Untold Story*, Relief Society of Tigray, 1985, plus other publications of REST; for Eritrea see J. Firebrace and S. Holland, *Never Kneel Down: Drought, Development, and Liberation in Eritrea*, Spokesman, 1984.
4. A. Aseffa, 'Ethiopian famine', unpublished paper, August 1986.
5. ibid., pp. 11–12.
6. ibid., passim.
7. Dervla Murphy, *In Ethiopia with a Mule*, London: John Murray, 1968.
8. International Bank for Reconstruction and Development, *World Development Report 1985*, Washington, DC: Oxford University Press, 1985, quoted in Simon Maxwell, *Food Aid Ethiopia: Disincentive Effects and Commercial Displacement*, Institute of Development Studies Discussion Paper 226, December 1986, p. 2.
9. John English, Jon Bennett, Bruce Dick, Caroline Fallon, *Tigray 1984: An Investigation*, Oxford: Oxfam, January 1984, p. 12.
10. Aseffa, op. cit.
11. Relief and Rehabilitation Commission (RRC), *The Challenges of Drought: Ethiopia's Decade of Struggle in Relief and Rehabilitation*, RRC: Addis Ababa, 1985, p. 55.
12. Richard Pankhurst (1961), quoted in Bahru Zewde, 'A historical outline of famine in Ethiopia', in *Drought and Famine in Ethiopia*, London: International African Institute, April 1976, p. 54.
13. Abdul Mejid Hussein, 'The political economy of the famine in Ethiopia', in International African Institute, op. cit., p. 14.
14. ibid, p. 12.
15. Mesfin Wolde Mariam, *Rural Vulnerability to Famine in Ethiopia: 1958–1977*, Vikas Publishing House in association with Addis Ababa University, 1984, p. 11.
16. Wolde Mariam, op. cit., p. 18.
17. James Bruce (1790), quoted in Bahru Zewde, op. cit., p. 55.
18. RRC, op. cit., pp. 58–66.
19. Wolde Mariam, op. cit., pp. 35–9.
20. ibid., p. 63.

21. Nigel Walsh (1975), quoted in *Lessons to be Learned: Drought and Famine in Ethiopia*, Oxford: Oxfam Public Affairs Unit, 1984.
22. Gopalakrishna Kumar, 'Ethiopian famines 1973–85: a case study', Balliol College Oxford, 1986, p. 8.
23. Glynn Flood, 'Nomadism and its future: the Afar', in International African Institute, op. cit., pp. 64–6; see also Hussein, op. cit., pp. 19–20.
24. Ministry of Agriculture (1973), quoted in Hussein, op. cit., pp. 23–4.
25. RRC, op. cit., p. 84.
26. Kumar, op. cit., p. 10.
27. Wolde Mariam, op. cit., p. 58.
28. A.K. Sen, *Poverty and Famines: An Essay on Entitlement and Deprivation*, Oxford: Clarendon Press, 1981.
29. RRC, op. cit., p. 91.
30. Wolde Mariam, op. cit., p. 4.
31. English *et al.*, op. cit., p. 9.
32. Guido Gryseels and Samuel Jutzi, *Regenerating Farming Systems after Drought*, ILCA's Ox/Seed project, 1985 results, International Livestock Centre for Africa, July 1986.
33. Aseffa, op. cit.
34. RRC, op. cit., p. 23.
35. Maxwell, op. cit.
36. International Bank for Reconstruction and Development, op. cit., tables 1 and 2.
37. Graham Hancock, *Ethiopia: The Challenge of Hunger*, London: Gollancz, 1985, p. 35.
38. ibid., p. 52.
39. ibid, p. 54.
40. René Lefort, *Ethiopia: A Heretical Revolution?* London: Zed Press, 1983.
41. Oxfam, *Tigray 1984*, op. cit., p. 11.
42. Peter Gill, *A Year in the Death of African Politics: Bureaucracy and the Famine*, Paladin Grafton Books, 1986, pp. 17–18.
43. ibid., pp. 19–21.
44. ibid., p. 23.
45. RRC, op. cit., p. 171.
46. Oxfam, *Tigray 1984*, op. cit., p. 15.
47. Gill, op. cit., p. 26.
48. RRC, op. cit., p. 172.
49. Hancock, op. cit., p. 66.
50. Gill, op. cit., pp. 17–18.
51. Food and Agriculture Organisation of the United Nations, *Ethiopia Report of the FAO/WFP Mission Assessment of the Food and Agriculture Situation*, Rome: United Nations, March 1984, p. 2.
52. ibid.
53. Gill, op. cit., p. 36.
54. RRC, op. cit., p. 174.

55. Michael Buerk (1984), quoted in Kumar, op. cit., p. 23.
56. Aseffa, op. cit., p. 6.
57. Aseffa, op. cit.
58. Unpublished figures from United Nations.
59. Clay and Holcomb, op. cit.
60. Concern's programme in Jarso and Ketto settlements, 'Evaluation team report', Nov./Dec., 1986.
61. Clay and Holcomb, op. cit., p. 195.
62. R. Dudley Edwards and T. Desmond Williams (eds), *The Great Famine* : Studies in Irish History 1845–52, New York, New York University Press, 1956, p. 1.

5
Case Studies of Famine Prevention: Botswana, Bangladesh, and Gujarat (India)

The famines in Ethiopia, Sudan, Mali, Chad, and Mozambique in 1984–5 attracted massive international publicity. Many other countries frequently suffer conditions of drought or floods which could lead to famine were it not for reasonably prompt and effective preventive and relief action by governments.

The three case studies in this chapter are all of countries which in the mid-1980s experienced floods (Bangladesh) or drought (Botswana and the Indian state of Gujarat), but which were able to reduce their impact considerably by comparison with previous occasions. The purpose of this chapter is to show why their performance in famine prevention has improved. Bangladesh, Gujarat, and Botswana are all different in climate, population density, and history. The relief systems of Bangladesh and Gujarat originate in the Famine Code established in India in the late nineteenth century (and adopted in colonial Sudan); but that of Botswana has been built up from very rudimentary beginnings at Independence in 1966.

A number of common characteristics emerge from the experience of Bangladesh, Gujarat, and Botswana which have improved their ability to prevent famine:

1. All have experienced political stability in recent years, as well as budgetary growth (from domestic and/or aid sources) and improvements in infrastructure and communications.
2. All have given an increased priority to famine prevention after previous harsh experience. This includes reorganizing and giving higher ministerial ranking to the relief administration.
3. All have geared up and diversified their early-warning systems,

through better and more frequent collection of crop forecast data, prices, and availability of essentials and indicators of hardship in particular areas.

4. All still rely basically on administrative means (agricultural and meteorological reports from local officers and statistics from clinics on nutritional status) rather than remote sensing for early-warning information.

5. All have improved their ability to react quickly and massively when disaster is imminent by intervening to support incomes through rapid deployment of labour-intensive public works projects, transfers of food and cash to those unable to work, and feeding schemes in schools and clinics. The achievement of Gujarat in the 1985–6 drought in also quickly mobilizing fodder from surplus areas in the state and so preventing an increase in livestock deaths, or collapse of livestock prices, was not matched by either Bangladesh or Botswana, where widespread stock losses occurred.

6. All have improved their ability to target their interventions to areas and people most in need – primarily the result of the improved early-warning system.

7. Finally, they are all states in which the long-term vulnerability of the population to famine has apparently not decreased. All are faced with the question of whether or not emergency relief activities are to become semi-permanent slack-season development activities.

DROUGHT-RELIEF PROGRAMMES IN BOTSWANA
RICHARD MORGAN

The country

Botswana's population is small in relation to its size: some 1.1 million people occupy 582,000 sq. km in the Southern African plateau. Due to shortages of surface water over much of the country, settlement is highly concentrated along the eastern belt, bordering South Africa and Zimbabwe, with only some 20 per cent of the people resident in the Kalahari Desert and the Okavango swamplands to the north. Villages are large in African terms, again due to water scarcity, although this is now alleviated by deep borehole drilling.

Economic activities permitted by the semi-arid climate have been limited in the past to the hunting of wild game and gathering of wild foods; small-scale cultivation, at very low yields, of cereals and pulses; and extensive grazing of cattle around seasonal water sources. Early in the present century, opportunities for employment in the South African mining sector drew a large proportion of the adult males away from the subsistence economy and left the rural areas heavily dependent on income remittances. Since Independence in 1966, the domestic mining sector has also become an important employer and has accelerated urbanization (currently about 19 per cent of the population are urbanized). A combination of borehole technology, favourable weather conditions, and the opening of lucrative European markets to Botswana beef provoked large increases in cattle numbers and the expansion of grazing into much of the western semi-desert areas during the 1970s. The benefits of these two expansions were, however, socially highly concentrated within the rural areas.

The diamond and beef export boom of the last two decades has brought national prosperity, and in particular the achievement of a favourable budgetary position and high levels of foreign exchange reserves. Bolstered by inflows of foreign aid, the Botswana government's development programme during the 1970s emphasized the provision of social services – schools, health facilities, and village water supplies – to rural communities; improvement of roads and other communications between the main settlements was also much in evidence. At the same time, however, the arable farming sector was stagnant, although the numbers of people dependent thereon continued to grow; and the expansion of the commercial livestock sector, mainly benefiting a minority of cattle-owners, tended to reduce opportunities for poor people to supplement their incomes through hunting and gathering. The benefits of national economic expansion through Botswana's incorporation into international commodity markets was not, therefore, reflected in increased household incomes for the majority of its people, except indirectly through social service provision and remittances. Those reliant on arable farming for part of their incomes remained highly vulnerable to climatic variations.

The incidence and effects of drought in Botswana

Botswana is one of the most drought-prone countries in Africa and is presently in the middle of a prolonged period of rainfall deficiency (occurring in the seasons from 1981/2 to 1985/6). Similar periods were encountered in the 1940s and 1960s, and Independence was achieved at a time when large numbers of cattle were dying and much of the population was reliant in part on food relief and Food-for-Work programmes. The present drought cycle, which began in 1978/9 following several high rainfall years, has reduced crop production from a range of between 40,000 to 80,000 tonnes per year to below 20,000 tonnes, leading to widespread income losses. Similarly, cattle numbers have fallen from a peak of some 2.97 million in 1982 to an estimated 2.4 million in 1985. Cattle losses have been concentrated dramatically among the smaller herds, the owners of which depend on communally used water sources, around which overgrazing is intense during drought. As a result, many poorer families have been deprived of their stores of wealth, and sources of draught power, occasional income, milk, and meat. Such losses have serious implications for nutrition levels among these families, as well as for their longer-term productivity in small scale farming activities.

Botswana's Ministry of Health introduced a nutritional surveillance system through its rural clinics in 1978, measuring the comparative weights and ages of children under 5 years old according to the Harvard Standard. As the sample grew in size, it was apparent that seasonal undernutrition (i.e. scores below 80 per cent of expected weight for age) was experienced by some 25 per cent of such children in normal years; this percentage increased to a peak of 31 per cent in the drought years of the early 1980s, and showed less seasonal variation. The surveillance system indicates that drought conditions in Botswana exacerbate an already chronic problem of undernutrition; these effects are greatest in remoter areas, but can be discerned in much of the rest of the country as well. A further consequence of drought has been a significant growth in the previously small numbers of adults, particularly older people, regarded by their communities as destitute and referred to the local district and town councils for assistance.

The government's response to drought

The alleviation of the immediate effects of drought has been a central concern of the national government since Independence. This derives from a number of factors; the close personal links of bureaucrats and politicians with the rural areas through their families and their cattle-holdings; the prominence of the experience of drought in Tswana traditional culture; and the reliance of the democratically elected ruling party on rural votes for its parliamentary majority since 1965. National experience in dealing with the consequences of drought provides a context for the conservative financial policies of the government and high levels of preference for investment and saving, whether in social infrastructure, cattle-holdings, or financial reserves, rather than spending for current consumption.

Despite this, the sudden ending of the sequence of 'good' years by the worst season on record in 1978/9 took the country very much by surprise. National structures had not been put in place to react quickly to such a situation, although the financial position of the country was much improved. Committee structures, lines of responsibility, and relief programmes were hastily designed, but due to lack of experience, particularly at local government levels, the impact was small and very late. Although the distribution of food was the main measure undertaken, lack of mobilization capacity prevented commodities, largely donor-supplied, from reaching recipients until shortly before the subsequent harvest. Fortunately, rural communities were able to withstand the one-year drought, and the following rains were good in most of the country.

The government recognized its failings and commissioned a thorough evaluation of the 1979/80 relief programme, which was conducted by consultants and discussed at a national seminar of officials. Subsequently, a number of basic decisions were taken at political level, including the recognition of temporary employment creation (which had been piloted on a small scale from 1978 onwards) as an integral part of the response to drought, and the restructuring of the food distribution agency within the Ministry of Local Government. It was also decided to use exclusively the schools and health facilities as distribution channels for food (except in the most isolated areas, where a special logistical effort was required), rather than the traditional village authorities, which had proved unequal to the task. Finally, the temporarily established Interministerial Drought Committee (IMDC) was confirmed as having national responsibility for the monitoring of measures in times of emergency. The co-ordination of the Committee is undertaken by the senior Ministry of Finance and Development Planning, which has easy access to domestic budgetary resources when required, and primary responsibility for contact with international donor agencies, including those assisting with food aid. Local authorities in the ten administrative districts were mandated to establish their own drought committees, under the direction of the IMDC.

When drought conditions returned in early 1982, the lead time necessary for decision-making had been much reduced. The relief programme, while initially concentrating on the provision of food supplies to 'vulnerable groups' at nutritional risk, developed subsequently into a multi-faceted effort. While those too young, or otherwise unable to work, are eligible to collect supplementary rations from health facilities (the uptake of which is very high), temporary work opportunities are offered at a basic wage during the agricultural slack season to the able-bodied. Jobs are created through the selection of low-technology construction projects by Village Development Committees, for which hand tools were provided and wages paid on a monthly basis by the district councils. Participation is rationed by queue, through a monthly turnover system, and has been expanded to some 70,000 people per year. A minority of the projects chosen are related to the agricultural sector (e.g. building of small-scale irrigation systems, clearing of firebreaks on grazing land), but most are complementary to the government's own programmes of social service provision. About 10 per cent of the jobs provided are in the primary schools, involving the hand-processing and preparation of sorghum for school meals. The overwhelming majority of job-seekers are women.

A third element of the relief programme is measures undertaken through the extension services of the Ministry of Agriculture to protect the capacity of small farmers to continue their agricultural operations following the end of drought. These programmes also have a broad remit, and include the provision of maize and sorghum seed to all households; grants for the clearance of fields prior to ploughing; and the subsidization of the hiring of draught power, where this is available, by households which have no animals of their own. Such measures incorporate aspects of the post-drought recovery programmes run in other parts of Africa following drought and famines, but are notable for being implemented simultaneously with the more strictly relief-orientated measures.

Effectiveness of Botswana's relief effort

The drought-relief programme has won plaudits from many observers for being among the most effective in Africa. Certainly, there is no evidence to suggest that starvation has occurred in any part of the country during the five years of drought since 1981, despite the cumulative effects on the rural economy. Since 1984, the nutritional surveillance system has indicated an improvement in levels of child nutrition, with the percentages of children found to be underweight falling to below 25 per cent in most of the country. Severe malnutrition has been limited to under 2 per cent of children and has been stable throughout the drought. This condition is thought to be related primarily to conditions in the family rather than climatic factors.

The agricultural programmes also appear to have achieved some success. The already high rates of rural-to-urban migration have not been greatly increased by the drought. In those parts of the country which did experience fair rains, a limited recovery in food production was apparent, particularly during 1985.

These limited successes during a period of severe stress on rural households can be attributed to a combination of organizational and strategic factors. The management of food relief, as well as of seasonal employment creation for large numbers of people, has been conducted entirely through the usage and strengthening of existing institutional structures, from national to village level. National planners have used existing data collection and information systems; transport has been reallocated from government departments for the trucking of food, water, and seed; local council workers have been extensively hired out to the strong private sector already existing in the country for the milling, storage, and transportation of food. In some respects, the relief effort has lacked professionalism but has been carried along initially through a sense of national participation and latterly through the build-up of basic experience from on-the-job training. The attention paid to monitoring, and regular national–district dialogue, has prevented mistakes from becoming too costly.

Certain preconditions for the success of the relief programme can be identified on this basis. The first is that institutional structures existed before the drought in functional form, whose capacities could be diverted to the extent necessary. This condition resulted from the build-up of social service infrastructure and from the decentralization of government responsibilities begun in the 1970s. The second is that complementary capacity existed in non-government sectors (the private sector, voluntary agencies, and donors), which could be mobilized as needed and to which the government was prepared to turn. The third is that domestic resources existed in the form of a stable financial situation, food importation capacity, and budgetary reserves, which could be employed in advance of the receipt of international assistance, and to cover those parts of the relief effort which donors were not prepared to fund. This applied not only to local transport costs but to almost the entire public works programme, which alone forms some 5 per cent of government spending from its own resources.

In terms of programme design, the relief effort embodies two important strategic conditions. First, the fairly strict selection procedures for beneficiaries of feeding programmes at clinics in operation before the drought were suspended, and supplementary rations were automatically made available to all applicants falling into nutritionally 'at risk' categories. This procedure has been expensive in terms of food and transport costs, but has made the programme simpler to administer, and, in providing a blanket coverage for children, pregnant and nursing mothers, has avoided the consequences of mistakes in selection.

Second, and most fundamental, the nature of the rural economy in Botswana has dictated the design of its relief programme. In a situation where food consumption is dependent as much on household income levels, including remittances and casual employment, as on crops grown, it has been necessary to find a means of preventing the collapse of rural incomes, and therefore purchasing power, during the drought. This has been done through the public works programmes, the scale of which has been related to the economic costs of the drought, and within which the basic wage has been increased periodically to account for inflation. This strategy of bolstering household demand has also prevented many rural traders, supplying food and basic consumer goods to agricultural communities, from going out of business.

Applicability of Botswana's relief programme design

The applicability of Botswana's experience in tackling drought is, however, constrained by these preconditions. Few countries in Africa combine democratically elected government, a reasonably efficient civil service at national and local levels, and considerable budgetary reserves, including importation capacity. Few possess the widespread private sector trading networks (including retail co-operatives) which can provide basic goods to workers on village-level public works. Additionally, Botswana's advantages in having a small population, concentrated in defined settlements, are not available to many countries seeking to reach target populations through food distribution.

None the less, the need for viable commercial structures to provide consumer goods in rural areas is increasingly recognized in much of Africa, and the use of labour-intensive works programmes to create a demand to support them, as well as to maintain purchasing power at household level when this comes under stress, is of potential relevance to this. Zimbabwe, for example, has used such an instrument from time to time, on similar lines to Botswana's programmes, in areas of localized drought. Lesotho has concentrated on Food-for-Work programmes in the past, but is increasingly moving towards forms of payment which will permit agricultural investment by seasonal workers, including the distribution of tools and seeds. In so far as Botswana's implementation capacity is enhanced by its use of early-warning indicators (including rainfall, soil moisture, crop forecasts, and nutritional surveillance), and by the maintenance of a small national grain reserve by the Agricultural Marketing Board to supplement food importation capacity in the short term, it is developing experience already recognized as being of relevance to the food security sector by the member states of the Southern African Development Coordination Conference (SADCC). In other respects, however, its response capacity is closely related to its high rate of overall economic growth since independence.

The limitations of Botswana's approach

A critical assessment of Botswana's experience has to recognize that, while much has been done to alleviate short-term drought effects, relatively little attention has been paid to reducing the underlying vulnerability of rural families to drought, or to mitigation of the longer-term consequences for them of the current drought. Lack of security of rural incomes is caused primarily by the extremely risky nature of the rain-fed arable farming systems on which many rural people will depend, and by the lack of seasonal or permanent job opportunities. Employment shortages can be tackled to some extent through public works, but at the price of rural dependence on the public sector and large public works expenditure. Given the low productivity of many projects, and the inability of the resource base to provide a context for most of them to become self-supporting or permanent income-earners, this outcome of dependence seems unavoidable. However, little has been done to identify and experiment with the limited opportunities that do exist, both to make short-term projects provide capital for sustained production and to give incentives to wage-earners to invest part of their incomes. Similarly, possibilities for the provision of skills and training in the course of public works projects have not been formally explored.

The economic base of the Botswana peasantry in off-farm employment and in low-yielding and high-risk arable farming in turn implies two approaches to reducing vulnerability, neither of them fully developed so far. Research on the existing farming systems, including investigation of alternative forms of draught power, use of drought-resistant crops and seed varieties, moisture-conserving planting techniques, and organic or chemical fertilizers, has commenced in recent years. But the results will not be available soon enough, nor will they be attractive enough to support a multi-year post-drought recovery programme in agriculture. The marginal position of many farmers, deepened by the drought and in particular by losses of draught animals, gives ample justification for increased levels of investment research and extension programmes around these problems.

The second possible approach to drought vulnerability, again conceivable in the near future as part of a rural recovery programme, is that of diversification of economic activities and farming systems. In part, this entails the regaining by many households of their positions in the livestock-farming economy, where greater economic returns are available; some form of state-assisted restocking is required for this. In addition, location-specific production is possible on the basis of resource investigation and exploitation: small-scale irrigation and forestry in the northern areas: wildlife utilization in the west; small-scale industrial development in the east. In contrast to measures to raise arable-sector productivity, this line of approach has received considerable government attention, and priority in budgetary outlays for a wide range of productive activities. The constraints on job – and income – creation in such sectors are imposed primarily by the smallness of the internal market, and a lack of comparative advantage in most products in major export markets. The increasing emphasis of the SADCC organization in facilitating trade between its members may provide new outlets for Tswana producers in future, assuming income levels rise in the area. But, faced with an annual growth in population of 3.3 per cent, the low rate of growth of formal-sector employment and lack of diversification in the rural economy indicate that reliance on small-scale farming as a primary source of food and income cannot be avoided. This in turn suggests that the maintenance of organizational and resource capacity for rapid response to the occurrence of droughts must remain a national priority for Botswana.

DROUGHT RELIEF AND DROUGHT-PROOFING IN THE STATE OF GUJARAT, INDIA

MICHAEL HUBBARD

India has long experience of recurrent famine, recorded back to the third century. Gujarat, on the north-west seaboard, is among the many regions of India which has relied substantially for its income, and still does, on the coming of the monsoon rains in the *kharib* (summer) season, June through September. Seventy per cent of the population earns its income from agriculture. If the monsoon fails, not only do rain-fed summer crops fail but groundwater levels fall, wells dry, and the *rabi* (winter) crops fare badly. Rainfall variation within the state and between years is considerable.

In most of the past twenty years a situation of scarcity through drought has been declared in one or more of the state's nineteen districts. The last two major droughts occured in 1975/6 and 1985/6. By contrast with the more distant past, when each major scarcity brought widespread deaths from starvation, hardship has now been reduced considerably as a result of the development and refinement during the past hundred years of an administrative system to counter scarcity.

Gujarat is regarded as one of the more wealthy and progressive states of India. As such its present policies in relief administration are probably more comprehensive and better administered than those of many other Indian states, such as the poorer and less-developed states of the north-east (e.g. Orissa and Bihar). The experience of Gujarat in drought-relief administration is that of a state with a large population of drought-vulnerable people: 22 per cent are agricultural labourers, the most vulnerable group; many of the farmers work small or marginal units (no irrigation potential). Both of these groups are poor, even in a year of adequate rainfall. Adequate relief management under these circumstances is a large-scale operation with a major challenge being to link the effort to investments which will reduce drought vulnerability in the long run.

Evolution of policy

Prior to the late nineteenth century no policy existed for famine prevention in British India, nor in the princely states. Food distribution, opening of relief works, and exemption from land tax took place in an *ad hoc* manner. Each famine was an isolated and unexpected phenomenon with which the government dealt hesitatingly and uncertainly, groping in the dark for some principle (Loveday 1914: 39).

Until the end of the eighteenth century 'the position of the British in India was not such as either to create any general obligation to give relief or supply sufficient means of affording it, according to the Famine Commission of 1880 (Dodwell 1929: 296). While the subsequent entrenchment of British administration created greater responsibility for famine relief, it was the shock of the Indian Mutiny of 1857 which produced a change in policy in favour of much greater investment in communications, infrastructure, and administrative training, as well as in agricultural and industrial production (Government of Gujarat 1976: 72).

In a supplement to the *Gazette of the Government of India* in September 1868, the measures to be taken for the relief of the starving in times of drought were codified for the first time; it was recognized that 'our famines are rather famines of work than of food'. A Famine Commission first sat in 1880, again in 1898 and 1901, and the Famine Codes of 1883 were revised repeatedly.

The principle on which the Famine Codes were based was that government had to ensure that there would be no loss of life during scarcities. 'The old principle that the public would be responsible for the relief of the helpless and infirm was entirely abandoned' (ibid.: 72).

The Codes provided detailed rules to guide the local administration in times of famine.

These included the establishment of a system of continuous flow of information, from every local area to the Provincial Government bearing on the state of scarcity or famine, the type and nature of relief works and the scale of wages paid, the organisation and gratuitous relief and the establishment of the system of village inspections, suspension of land revenue and grant of tagavi [i.e. seed] loans, relaxation of the forest laws for the duration of the famine and protection of cattle. The district was to be the unit of administration for famine relief.

(ibid.: 75)

While the early Famine Codes established the principle of government responsibility to prevent loss of life and preserve livestock, and laid down administrative systems for the purpose, administrative practice was not always equal to the task – most tragically in the case of the great famine in Bengal in 1943–5. Similarly, though much less tragically, was the scarcity in Maharashtra and Mysore in the drought of 1952–3 (documented by Dandekar and Pethe 1972), which developed to an advanced stage of famine-distress sales of livestock, with total livestock holdings down between 20 and 30 per cent, large-scale migration of people over 20 to 30 miles, emergency relief camps (*annachatras*) set up with temporary accommodation and feeding – the sort of conditions familiar from the recent African famines, which indicate that relief has been too little and/or too late.

Thus the priority was to improve administrative performance, and the later reforms of the Famine Code (as in the Bombay state *Scarcity Manual* published in the 1950s and the current *Gujarat Relief Manual*) are built not so much on altered principles, though there has been some elaboration of these (e.g. the commitment during scarcity to supply work to everyone who needs it not further than 5km from their home, as on the need to detail administrative responsibility and to quicken response. Behind the administrative reform lies financial reform: funds are provided by central government subject only to the restriction that allocations do not exceed the state's allocation for relief in the Five-Year Plan. Funds for relief have increased substantially, providing the basis for the claim by Gujarat relief officials that their relief programmes are now 'need based', with the state's objective in times of scarcity now being not simply that no one should die of starvation but also including prevention of physical deterioration and destitution of people, and enabling them to resume ordinary pursuits of life on return of better times (Government of Gujarat 1976: 83).

Implementation of relief policy

With the basis for famine prevention administration in India being the district, the administrative head of the district (the deputy commissioner, or collector) is at the centre of operations in every state, the difference being how much decision-making autonomy the collector is given by the state government. In Gujarat the collector is responsible for all implementation decisions except the initial declaration of scarcity, which must be done by the state government, although the collector initiates overall direction and monitoring and release of funds. Central government provides the bulk of the financial assistance. In Gujarat the relief department in the state government has been strengthened in recent years, from being an *ad hoc* office in the revenue department to a permanent office with a director and commissioner (during scarcities) and a 'collective memory' of statistics. It is also the 'control cell' for forecasting impending drought in the state.

A picture of how relief is administered can be built up by following through the steps taken each year to monitor the monsoon and relieve scarcities developing from inadequate rain in any area – as occurred severely in several districts in Gujarat in 1985–6.

Early warning

The relief department starts receiving daily rainfall reports from each of the 184 *talukas* (development blocks, containing about 100,000 people) on 1 June, the earliest date at which the monsoon can be expected to start sweeping up over the sub-continent. It is the personal responsibility of the *tahsildar* of each *taluka* (chair of the *taluka panchayat* – the block-level council) to phone this figure through daily to the relief office. By aggregating the data and comparing it with previous years the probability of drought developing in any part of the state is assessed (Koshy 1986: 3).

Similarly, a flood-control cell is run by the state government supplied with daily data on reservoir levels, on the basis of which it is possible to assess when to cut down or ban the use of a reservoir's water for irrigation so as to preserve it for drinking purposes (as was done in October 1985).

As the monsoon months begin the district commissioner must be kept informed of the likely state in all parts of the district of the *kharib* (monsoon season) crops, the levels of reservoirs, and availability of drinking water, the condition of the people, the extent of food grain and fodder supply, the state of exports and imports of food grains and fodder into the district, the trend of wages, opportunities for employment on agricultural and other operations, mortality statistics, unusual move-ment of labour in search of employment, and the state of crime, particularly petty grain thefts (Government of Gujarat 1982: 12).

If the district commissioner decides it is necessary to open relief works to generate employment or to distribute gratuitous relief, then he must report fully on all these to the state government and recommend suitable relief measures where the threat is serious. Test works to assess the demand by people for employment may be opened and managed departmentally (e.g. irrigation or forestry departments); if a test work attracts only a small number of people from the locality then it might be concluded that relief employment is not required.

Although the monsoon may last into October, it is clear by September whether there will be shortages and how acute they will be. In 1985 the early-warning system had begun to indicate likely distress by the end of July, and a full-scale 'drought alert' was issued by the beginning of September (Koshy 1986: 7). In the event, 127 of the 184 *talukas* of the state had received significantly deficient rainfall by early October, and test works and fodder distribution began even before the end of the monsoon. Late rains in October in some areas gave a better prospect for the *rabi* (winter) crops, but the state still faced the worst scarcity situation in ten years.

As soon as it was clear that a major scarcity threatened, a cabinet sub-committee on relief was set up in the state government, meeting weekly, chaired by the finance minister and serviced by the relief commissioner. At the same time, a state-level relief committee (including members of the opposition parties and voluntary organizations) and district and *taluka*-level relief committees were constituted, and each cabinet minister was assigned special responsibility for one or two districts. The immediate task was to prepare the District Master Plans.

The District Master Plan

The District Master Plan for relief is the centrepiece of relief strategy. It is village based and prepared in consultation with the local *panchayats*, who are also responsible for much of the implementation, and particularly for bringing further cases of need to the attention of the district authorities. The plan is prepared in the district and approved and funded by the state; given the urgency of getting relief operations going it is prepared and finalized within days. Provisions of employment, drinking water, fodder, and distribution of 'gratuitous relief' to those unable to work, are the four activities covered by the Master Plan.

Employment generation through public works

Each affected *taluka* is divided into village groups of four or five villages, the groups or circles (as they are known) becoming the basic administrative unit for public works, their size dictated by the policy of providing employment within 5km of home to those wanting it.

A survey within each is done in order to establish the numbers that need to be provided with work each month until the rains begin again in June of the following year. The estimate made is based mainly on the numbers enrolling on works in previous droughts, and on the numbers of small and marginal farmers and agricultural labourers. A rule of thumb of providing relief employment for 10 per cent of the population is sometimes followed. Interdepartmental consultations take place, involving mainly the irrigation, forestry and public works departments, to establish the availability of already budgeted works (e.g. under the national Rural Employment Programme, Rural Landless Employment Guarantee Scheme, or departmental projects) and what the shortfall in person days of work will be: to be made up by initiating new projects taken from the 'shelf' of projects which each department in the *taluka* must have to hand, ready for quick implementation in consultation with the *gram panchayat* (village council).

There is a ranking of priorities of relief projects: most critical is that they can offer employment to large numbers of unskilled people with as little materials, tools, and machinery costs as possible. Works which will help people through future droughts are preferred, including 'every known project of irrigation' (mainly village tanks, percolation tanks to raise the water table, and check dams on rivers), particularly those which can be completed or brought to a safe stage within the period of the relief employment. Works which improve utilization of ground water are preferred as this is a more dependable source. Contour bunding works for rainwater harvesting and soil conservation include gully plugging, stream bunding, terracing, and land levelling. Forest works (tree planting mainly) are preferred over road works, and the breaking and stacking of gravel for existing roads is preferred over starting of new roads, 'the likelihood of maintenance of which after scarcity is doubtful' (Government of Gujarat 1982: 5).

The organization of work, payment, eating, and health facilities at each works site occupies much of the space of the *Gujarat Relief Manual*. Rates of pay are laid down at the state level, determined by the quantity of work done by the individual, the type of task (e.g., in the case of earthworks, the quality of soil worked with and the distance and height carried), with exceptional additional payment being made according to the number of dependants of the worker in cases of extreme hardship. There is a six-day week, with a rest-day allowance payable.

The rates are tied to the retail price of food grains and are designed to be 'the lowest amount sufficient to maintain healthy persons in health' (ibid.: 43). Workers are organized into gangs about eight strong, with weaker persons put into separate gangs which may have a smaller quota of work to perform daily in order to earn the Rs11 wage, as is also the case for new workers until they acquire facility at the task. There are higher rates of pay for those involved in heavier work (e.g. digging) than for carrying, or for serving food and water to the workers (the job reserved for the weakest people). No one under 14 years of age may be employed on relief works.

The amount of work done by each person in the gang is measured daily and payment made weekly in the presence of the *sarpanche* (village council administrator) or representative. In order to minimize the possibility of fraud, the measuring and payment are now carried out by different departments and are subject to surprise checks by senior officials of the *taluka* district.

Pay may be entirely in cash or a combination of cash and food grain. On one project visited in Rajkot district, pay was Rs5.5 plus 3kg of wheat per day. With pay rates tied to grain prices and a market for grain at hand, the medium of payment should not make a great difference; indeed, in the neighbouring state of Rajasthan, where relief payment in the drought of 1985–6 was wholly in food grain but the price of grain in the Fair Price shops was below the market price, workers reportedly preferred to receive wages in kind since they yield more than the relief wage in cash when sold on the open market.

Whatever the case, the pay system relies ultimately on essential goods being available at expected prices. In India this depends on whether the government distribution system for essentials through the Fair Price shops (at which every one can buy to the extent of their ration cards) is operating at all well. In Gujarat the system has a good reputation (though the *Gujarat Relief Manual* cautions officers in charge of relief works to ensure that a Fair Price shop is available close to the site and is open on pay day every week). In Rajasthan the erratic opening hours of the Fair Price shops and their tendency to carry only sugar resulted in the change of the relief payment system away from food grain vouchers redeemable at Fair Price shops to direct payment in wheat at the work site – with attendant difficulties in transport for both payers and receivers.

Relief works are kept going until there is no more demand for employment on them or until the rains come, when they are wound down (generally the end of July) so that people will be encouraged to return to their fields or to offer themselves as agricultural labourers. I was told that this does not cause hardship since alternative employment was available and that if the works were not closed, many would not go back to agriculture since they can earn more on the relief works. In some cases the winding down of works is accompanied by reduction in the wage rate in order to encourage people to return to agriculture.

Drinking water

As the drought develops, the water table drops and surface water disappears, the district administration undertakes surveys to find possible drill and well sites, shortlists the problem villages, and starts drilling, digging, constructing pipelines, and running water tankers to the most deprived areas. The norm adopted for planning is that at least 9 litres of water per head per day should be available for the rural population. In 'at risk' areas the use of reservoir water for other than drinking purposes is prohibited during the scarcity. In the drought of 1985–6 cityl alcohol was used for the first time on the surface of some of the state's reservoirs to retard evaporation, apparently with some success (Koshy 1986: 18).

Fodder

The principle adopted by government is that there is a public responsibility for keeping alive the cattle belonging to the 'weaker sections' of the rural population (i.e. small and marginal farmers, agricultural labourers, and scheduled castes and tribes) by providing them with a minimum of fodder.

At an early stage the state government bans any export of grass and obliges all suppliers of grass to sell to government. Deputy commissioners estimate what the shortfall in fodder to eligible classes of cattle-owners in their districts will be. As Koshy describes, in the case of the 1985–6 drought, the projected shortfall was far above what could be supplied in the main grass-growing area, Valsad, in the south of the state (Koshy 1986: 14, 19). The shortfall was made up by the conversion by sugar factories of surplus bagasse into low-cost feed by steaming it in a digester and adding small quantities of urea and molasses. Second, an agreement was made with flour mills whereby half of their total production of wheat bran was to be purchased by the government. Third, the government entered the market for paddy and wheat straw. At the same time a massive road and rail transport operation was organized to move 80,000 tonnes of grass from Valsad in the south to the fodder-deficient areas in north and west Gujarat.

Voluntary organizations (mainly religious, *Gaushalas* and *Panjrapoles*) have played an important role in preventing cattle deaths, by setting up feeding camps for the cattle of people, and even of whole villages, that cannot feed them. For this they receive a government subsidy of about one-third of the feed costs, the rest coming from private sources (ibid.: 20). There are specific rules excluding support of old or economically useless animals – but it is difficult to see how this can be rigorously enforced.

Gratuitous relief

The distribution of gratuitous relief normally begins before the opening of relief works. People eligible are those who are disabled and have no one able or willing to support them, or those whose attendance on the disabled or young children is essential, and men and women 'of respectable birth' who risk starvation, for whom no suitable employment can be found on relief works, or who by custom are unable to appear in public.

Those requiring gratuitous relief are identified by the weekly village inspections carried out by the relief circle inspectors under the direction of the collector. During the weekly inspection the administrators ensure that adequate information is posted regarding the relief works available in the circle and maintain the list of people qualifying for gratuitous relief, each of whom receives a relief ticket in their own name. Half of the allowance is given in cash, the other half in grain, and is paid monthly. Regarding the stoppage of gratuitous relief the relief manual lays down that

> As the demand for agricultural labour increases, the gratuitous relief lists should be more carefully scrutinized, with a view to their reduction. When the earliest autumn crop is ripe in any tract, gratuitous relief shall be generally closed in it, whole villages being struck off the list at a time.
>
> (Government of Gurajat 1982: para. 160)

Assessment of relief organization in Gujarat

The performance of a relief effort is measurable firstly by results and only secondarily by the efficiency of resource use. All officials that I spoke to at district level and in the state capital were of the opinion that the relief effort in the drought of 1985–6 had been successful: there had reportedly been no starvation deaths recorded, virtually no migration of people (probably even less than normal in the case of young adults as more unskilled wage employment than usual was locally available), and no recorded stock losses or stock migration. The size of the relief effort was estimated to be the same as in the last major scarcity, in 1975–6.

The checks on the performance of the relief effort are internal (the numerous mechanisms detailed in the relief manual whereby senior officers are to visit work sites and villages to check the work of juniors), political (through the *panchayati raj* up to the legislative assembly of the state), and public (newspapers). The administration does not commission independent consultants to assess the overall impact of relief but did seek outside advice this year on one troublesome aspect, namely the delays of up to three or four weeks incurred in making payment to workers on relief works.

In visiting relief operations in Rajkot district and relief headquarters in the state capital, I encountered close interaction between officials and local and state-level politicians. 'Political careers are made and broken during scarcities', one senior officer told me, indicating that politicians compete with each other to be seen to be effective during the time of need, with opposition parties eager to show up any deficiencies in the effort. Similarly, voluntary organizations fulfil a watchdog function as well as performing practical work such as running cattle camps. The most noteworthy case I encountered was in Rajasthan, where a voluntary organization was in the process of suing the government for paying less than the national minimum wage at relief sites. Gujarat is one of the most literate of Indian states and local private newspapers abound; unlike Indian television, many newspapers seek out local controversy, and officials are aware that bungling will not necessarily remain an internal administrative matter.

Sen has surmized that it is the watchdog role of a relatively free and active press which has enabled India to be more successful than China in preventing famine since Independence (Sen 1983: 758). Whatever the merits of the comparison, my impression from Gujarat was that active local politics in the *panchayati raj* institutions plays as large a role, together with a civil service that has much technical competence at the local rural level (particularly strong in engineering, which means that plans for useful relief works can quickly reach the implementation stage) and is willing to put in greatly increased effort and hours during times of scarcity with no extra pay.

Finally, the entire relief effort is built on the ability to respond quickly, to move stocks of food, feed, and materials in time, to store and distribute them and account for them. Basic food distribution in India is heavily regulated and handled largely by the state – either directly by a government department or indirectly through a parastatal organization, as in the case of the Civil Supplies Corporation in Gujarat. During times of scarcity it is widely recognized that food supplies tend to gravitate towards the cities, and all countries during such times have tried to regulate food movements. Whatever the demerits of the close regulation of the marketing of basic foods during normal times, the pre-existence of a tolerably competent state distribution system is an advantage during scarcities.

It is widely recognized that India's major development failing has been the persistent poverty and undernutrition of much of its rural population, closely related to rapid population growth. Its vulnerability to famine during droughts is a result, and in this light the effectiveness of counter-famine administration might be seen as a response to failure of development policy, essential though it may be. The great challenge is to diversify and increase the incomes of the poor so that they become less drought vulnerable; it remains to be seen whether India's highly stratified society will be able to accomplish this task as well as it has (at least in the better administered states, such as Gujarat) managed to combat famine, largely through a very hierarchical administrative system.

FLOODS IN BANGLADESH, 1974 AND 1984: FROM FAMINE
TO FOOD-CRISIS MANAGEMENT[1]
EDWARD CLAY

Bangladesh, formerly East Pakistan, is one of the poorest countries in the world. It is flat, very densely populated, and has a long history of devastating floods.

In 1984 Bangladesh suffered monsoon flooding as severe as in 1974, the year of famine. The flood losses in production inevitably put severe strain on the food system. The strains were so severe as to be characterized as a 'food crisis', that is a situation in which government resorted to extraordinary measures. The seasonal losses in production for the three rice crops of the calendar year 1984, *boro*, *aus*, and *aman*, were at least 1 million tonnes.[2] As we now recognize, such a loss in production also results in a drastic reduction in employment for the rural landless, on a conservative estimate 25 million person-days. There were increases in severe cases of malnutrition amongst children and reports of forced sale of cattle on a large scale. These are circumstances which could have led to famine conditions. There has clearly been severe stress affecting millions of people, but despite this there was not a famine in 1984.

Why was there no repetition in 1984 of the famine of 1974? Several important institutional changes in the agricultural system, in the food system as managed through government intervention, and in the political system more generally, appear to have contributed to famine being averted.

Some suggest that the institutions of government (the Ministry of Food and the Ministry of Relief and Rehabilitation) are, in a purely administrative sense, functioning more efficiently than in 1974. Bangladesh had been independent for scarcely two years when the crisis broke in 1974, and administration had been severely disrupted by the war of independence. There was still considerable internal political instability verging on civil war. Returns for the 1974 census were falling off the backs of lorries. (There was a vigorous market in recycled paper.)

Arguably, there is still political instability. But there has been considerable institution building in the management of the food system. First, there are elements of an early-warning system in place operating more effectively than a decade ago. This is in part simply because of the normalization of the operations of government in comparison to 1974.

Second, there is now an accurate system for monitoring the operation of the food grain system reflected in the *Monthly Foodgrain Review* published by the Ministry of Food and the *Monthly Foodgrain Forecast* circulated more widely than the World Food Programme (WFP). The government review reports in detail on the food system outlook. Both publications report recent and projected movements of stocks through the system, the arrival of food grains, public distribution, internal procurement, storage position, and more recently, even food grain prices. Doubtless all the information is not entirely accurate, but the important thing is that it is believed and government and donors act on it. For those observing the current crisis in Sub-Saharan Africa, this strong informational basis for food system management and donor action stands in sharp contrast to the situation, for example, in Ethiopia.

Third, there is the hard-to-assess level of government commitment to containing price movements and maintaining distribution of food as well as preventing food-related stress on the poor getting out of hand. It is not that governments in Bangladesh were not always committed to preventing famine but that they now understand and have better information about the operation of the system. Perhaps also, in the aftermath of the events of 1974, higher priority is accorded to providing a rural food-security net and preventing the rundown of stocks to a level where government can no longer intervene effectively to maintain simultaneously urban and rural food security. This assessment is reflected in the reaction of government in 1984.

At the time of the earliest flooding in May and with an election in prospect, the martial law regime made up its mind to ensure an adequate pipeline of food so that it could maintain food system operation. By the end of May it had entered into deals with Thailand for large-scale commercial purchases. In the end, these purchases amounted to over 400,000 tonnes of rice. The government had reason to act. Stocks of rice were below 100,000 tonnes. But for some early arrivals of Canadian aid, stocks could have been at a very low level by November, and there was a December election in prospect. The point to note here is that the government was not prepared to gamble with food security. Then followed the floods of June and September. Some emergency food aid was committed by the WFP and the pipeline of assistance began to fill up.

A fourth area in which government behaviour is different from the mid-1970s is reflected in the way it manages the food system. In terms of offtake and procurement, the food system is now managed more closely in relation to price movements. The government has not sought to intervene as it did in 1974 and in earlier food crises of the Pakistan era through forced procurement to acquire grain, thereby actually disrupting commercial trade and probably forcing up prices. There are now open-market sales of grain, relatively small in relation to total public operations (less than 5 per cent offtake of wheat in the first three months of the agricultural year 1984–5), but nevertheless indicating a changing pattern of food system management.

Fifth, there are other large-scale interventions in place intended to contribute to rural food security as well as, year in year out, providing livelihoods for the rural poor. These include the Food-for-Work programme supported by the WFP, USAID through CARE and other major food donors. There is also the Vulnerable-Group Feeding programme established in the mid-1970s and which has more recently become a major part of public distribution (accounting, for example, for almost 10 per cent of wheat distribution in September 1984). This programme, notionally a supplementary feeding programme, is now widely accepted as providing an income transfer to poor rural households, targeted on women and children.

Sixth, in the management of interventions there is now some attempt to base the regional and seasonal distribution of allocations on poverty and disaster mapping. For example, the Food-for-Work programme in

1984 was expanded by more than 50 per cent in the apparently worst-affected areas. This notion of mapping goes back to the 1974 experience and research such as Bruce Currey's study of famine incidence.[3] The large-scale Food-for-Work programme is now used, as it has been since the late 1970s, as a flexible intervention that can be expanded considerably in a crisis. In current terminology there is now dry-season food-for-work, test-relief work for food in the wet season (a category carried over from the earlier famine code) and gratuitous relief.

Seventh, another significant change is the considerable build-up in storage capacity from something over 1 million tonnes in 1980 to 1.9 million tonnes in 1984. Storage is now widely spread throughout the country. All these changes add up to considerably improved capacity within the public distribution system to move and store large quantities of food. This storage capacity allows the government to handle large working stocks. Imports were estimated at some 2.8 million tonnes in the year July 1984 to July 1985 and arrivals were bunched. In a more favourable year perhaps 1 million tonnes might be procured and temporarily stored in the months after the main *aman* harvest in November–December. There is a capacity to intervene to contain seasonal price movements exacerbated by a combination of imports, local purchases, and massive distribution. The storage is not used for large-scale inter-year storage and buffer-stock operations.

There have been some other significant changes in the decade. The agricultural system is more diversified, and therefore resilient against natural disaster, because of the expansion in winter-irrigated *boro* rice and wheat production. Statistical analysis has suggested that the overall variability in crop production year on year is declining slightly as a result of changes associated with irrigation, flood control and changes in cropping patterns.

Another change is the improvement in relationships between government and donors. There is a local consultation group meeting to discuss the food system. The monthly statistical documentation already noted reflects a willingness to exchange information on the food system operation. In contrast to 1974, when political differences between government and donors, particularly the United States, intensified problems, no such problems exacerbated the 1984 crisis. The change in relationships is reflected also in the considerable extent to which food aid is now programmed in terms of multi-year commitments. There are multi-year commitments of both PL480 Title III and Canadian programme assistance. Food-for-work and Vulnerable-Group Feeding, both multi-annual projects, represent a larger proportion of total food aid.

There are some other important differences in the economic environment between 1974 and 1984. In 1974 the country was in a weak economic position after the war. The infrastructure of roads, rolling stock and storage was not fully rehabilitated. There were problems of industrial production, again related to rehabilitation (e.g. the breakdown of the sole factory producing nitrogenous fertilizer). Bangladesh now has greater purchasing power and credibility as a purchaser looking for credit agreements. This is particularly a consequence of the rehabilitation of the economy and the way in which Bangladesh has recently benefited from an improved foreign exchange situation, largely by worker remittances from overseas. The wider economic environment is less unfavourable. The world grain market conditions made it possible for Bangladesh to go out and purchase large quantities of rice in 1984.

How important is politics? There is unquestionably considerable corruption in Bangladesh now, as there was in 1974. Yet priority is accorded to food security. This was indicated when government just managed to weather the food crisis in 1979–80 brought on by drought. It was reflected again in measures taken in 1984. The ration system and project food aid interventions are leaky. Estimates of leakage vary widely. Few would put these at less than 30 per cent in circumstances when the Dhaka High Court is unwilling to consider cases of corruption and misappropriation of less than 30 per cent (a recent reported ruling). Evaluation suggests that less than 60 per cent of Vulnerable-Group Feeding reaches the intended beneficiaries. Nevertheless, the well-articulated public distribution system through which large quantities of grain are poured into rural areas, and on a considerably larger scale in periods of stress, dampens any possible effects of reduced food production on prices and helps to sustain livelihoods of many of the most vulnerable.

The scale of these changes is reflected in the Food-for-Work and Vulnerable-Group Feeding interventions. In 1984/5 it was planned that Food-for-Work would expand from 410,000 tonnes to something like 500,000 tonnes (an increase of 25 per cent). The bulk of the additional resources are focused on flood-affected districts. The nominal increase is possibly equivalent to the loss in person-days of employment. It is not, however, targeted on those areas affected in proportion to loss of employment. Such fine targeting is impracticable with existing assessment procedures and institutional mechanisms. The Vulnerable-Group Feeding programme is envisaged to expand from around 120,000 tonnes to some 200,000 tonnes, an increase of 70 per cent, involving a nominal 1.25 million mothers and their children. Officials in discussing the Vulnerable-Group Feeding programme describe this as an income transfer equivalent to perhaps 20 per cent of income for 2 million persons. Those interventions are on a considerable scale, and this is no mean achievement. This battery of interventions has succeeded in considerably dampening the impact of environmentally induced disaster and crop loss on the rural population and the economy more widely. The effect is sufficiently impressive for many to see such interventions as models, alongside the employment guarantee and Food-for-Work programmes in India, for what is needed in Africa.[4] And yet there is little reason for complacency.

The large-scale losses of cattle in 1984 are a reminder that other complementary interventions to prevent the erosion of productive capital are still not in place. Parallel to current interventions, at the very least some form of credit system that actually reaches small farmers is necessary to counteract the effect of environmental stress. Even though there was no famine in 1984, the process of impoverishment, which also inhibits long-run agricultural growth and, in a broader sense, rural development in Bangladesh, has been intensified once again by flood.

Notes and references

1. Edward Clay and the publishers would like to thank the editors of *Food Policy*, PO Box 63, Westbury House, Bury Street, Guildford, Surrey, GU2 5BH for their kind permission to reproduce this article, which first appeared as 'The 1974–1984 floods in Bangladesh: from famine to food crisis management', in *Food Policy*, vol. 10, no. 3, August 1985, pp. 202–6.
2. R. Montgomery, 'The floods of 1984 in historical context', Dhaka: USAID, 1984 (draft mimeo); Ministry of Agriculture, 'Comparison of 1984 and 1974 floods', Dhaka: MoA (draft mimeo); Government of Bangladesh, Ministry of Food, *Monthly Foodgrain Review*, various. The government estimates losses at around 1.5 million tonnes. However, other observers, including Montgomery, estimate losses at only 1 million tonnes. There are three, arguably four, seasonal rice crops in Bangladesh. The *aus* crop is planted in March and April and harvested in July and August, during the monsoon rains. Broadcast *aman* (deepwater) rice is planted from March to May and harvested from late October to December. *Aman* (the main crop) is transplanted between July and September and harvested from mid-November into December.

Boro (largely irrigated) is transplanted from November to February and harvested from April until June.
3. B. Currey, 'Mapping areas liable to famine in Bangladesh', PhD dissertation, Honolulu, University of Hawaii, 1979 (University Microfilms Order No. 8012253, Ann Arbor, Michigan).
4. See, for example, the contributions of R. Azad and T. Page to the Third IDS Food Aid Seminar, in E.J. Clay and E. Everitt (eds), 'Food aid and emergencies', Institute of Development Studies, University of Sussex, 1985 (draft mimeo).

PART II

POLICY ISSUES IN FAMINE PREVENTION

6

Emergency Measures for Food Security: How Relevant to Africa is the South Asian Model?

ASSESSMENT OF FOOD-ENTITLEMENT INTERVENTIONS IN SOUTH ASIA
EDWARD CLAY

The food policies of South Asian countries are characterized by massive interventions to provide food entitlements to potentially vulnerable groups and to other categories of consumers whose food security has a high political priority. There are broadly three categories of intervention:

1. Ration and other food-entitlement schemes which typically combine the dual objectives of assuring supplies (food security) and subsidized consumption (income transfers) to eligible groups.
2. Rural works/Food-for-Work which combine guarantee of employment (and therefore of food security) with developmental investment objectives.
3. Vulnerable-Group Feeding, typically mother and child and school feeding programmes which are directed to improving longer-term nutritional status and human resource development but which, in assuring some food transfer to vulnerable groups, have a clear food security function.

From the viewpoint of public policy these measures should not be considered as just separate sets of interventions which coincidentally contribute to food security, but as part of a food security net. Additional food was pumped into the food system through all these mechanisms in 1984 in order to counteract a threat of famine.

National food-entitlement programmes[1]

The ration system in Bangladesh, Pakistan, and Sri Lanka up to 1979; Fair Price shops in India; and more recently food stamps in Sri Lanka, combine food-security with food-subsidy objectives. These broad programmes all operate through a mixture of price controls and rationing. The extent of subsidization is restricted by limiting *ration eligibility* and by supply of products which are regarded as inferior items of diet. All of these programmes have been supported at different times by large-scale programme food aid. Recently, they have been subjected to close scrutiny as budgets have come under increasing pressure from national treasuries and the International Monetary Fund (IMF). The general conclusions that emerge from these evaluative studies and also parallel research on Egypt[2] and Indonesia are as follows.[3]

Programmes have had a significant positive impact on food consumption and the real incomes of poor people. The benefits of entitlement programmes are generally widely distributed. Such programmes are either explicitly targeted through rationing, as in Bangladesh and Pakistan, or effectively only reach urban populations. However, where the urban poor typically have low levels of food consumption this bias may have a positive aspect. But there are contrasts. The bread subsidy and rice rationing in Egypt and wheat subsidy in Sri Lanka prior to the 1977–9 reforms had general coverage, in contrast to the more selective targeting in Bangladesh and Pakistan, where eligibility criteria and administrative practice had the effect of excluding the poorest urban households.[4] The little research on food-entitlement programmes done so far for African countries suggests that their systems operate more like those in Bangladesh and Pakistan.[5]

The experience in these countries demonstrates effective and administratively least-cost ways in which commodity subsidy programmes can be focused on poorer households. The desirability factor is particularly important in determining the distributional impact of food subsidies. Where the subsidized distribution is handling less desirable basic food staples, an 'inferior' commodity which only poor people are prepared to accept, then the commodities are 'self-targeting'. By coincidence, wheat, the most widely available commodity, through trade and aid, represents an 'inferior staple, in much of the sub-continent. The exception is semi-arid central India, where wheat is preferred to the traditional staples, sorghum and pearl millet or even rice. Where imports and food aid in particular can only supply a 'superior' food, that is one which suits upper-class tastes (wheat and rice in most of Sub-Saharan Africa), the problem arises that it may be leaked unofficially towards these classes. One response may be to sell the superior food that is supplied by aid and use the proceeds to buy the inferior, self-targeting foods at full price from producers for subsidized or free distribution. This would then have limited risk of leakages. Clearly, such a desirable scheme would depend for its operation on local circumstances and government pricing policies. It also presumes that the self-targeting foods are domestically produced and predictably marketable on an adequate scale.

Supply manipulation and differential pricing policy for food staples may offer, on the evidence from these countries, some opportunity to get cheap calories to consumers. This then becomes an important component of food security. Aid can provide resources to sustain such a policy. Again, as experience in these countries illustrates, actual government practice and (looking to the future) capacity is a critical variable. Experience on the one side in Sri Lanka up to 1977, Kerala, and also Egypt, where assured supplies at subsidized prices were general even in rural areas, stands in contrast to the much less satisfactory experience in, for example, Bangladesh. Narrow targeting, particularly on urban populations, can increase the vulnerability of those outside the subsidized and stabilized system, who are likely to be the rural poor and those sections of the urban poor who face difficulties in establishing their entitlements under such systems. This was most obvious during the Bangladesh famine of 1974. The most vulnerable people were in remote or inaccessible areas and groups who were socially 'marginal'. The shift in Sri Lanka from general subsidy to targeted food stamp entitlement also appears to have resulted in practice in the exclusion of some highly vulnerable groups in the population, such as newly established poor urban households and workers on the tea estates.

The cost-effectiveness of broader subsidy programmes continues to be an area of controversy. On one side there is a substantial body of concern that these programmes create severe budgetary and balance-of-payment problems and weaken incentives for agricultural producers. But studies by Scandizzo and others at the World Bank found that such subsidy programmes can nevertheless be cost-effective.[6]

The nutritional impact of subsidy programmes is another area of controversy. Some studies have indicated that food subsidies are an effective way to improve nutritional status of low-income households and vulnerable groups, for example children in Kerala. But such findings are in turn contested by others pointing to the possible influence of interacting factors such as public health provision.[7] Well-documented experience notably in Egypt and Sri Lanka (before 1977) show that such programmes, if not a sufficient condition for a significant and sustainable improvement in the nutritional status of vulnerable groups, nevertheless contribute along with provision of other basic needs to making a significant impact on measures of well-being such as mortality rates and the incidence of malnutrition in these countries. As a realistic component of a food security strategy, the question also remains whether governments are able to sustain these programmes when they come under pressure in periods of food crisis. The weakness of an administered distribution system is that when food is scarce and logistical capacity is under pressure the most vulnerable may have lowest political priority. Powerful and independent media, as in India, can play an important balancing role. The separate organization of the support for interventions directed primarily at vulnerable groups, Food-for-Work and Vulnerable-Group Feeding, can also play an important role in redressing the balance.

Labour-intensive public works and asset formation[8]

Public works have a long history as a food security measure in times of distress and famine. Means of survival are provided for the destitute, and households may be spared the necessity of enforced sale of assets to provide short-term consumption.

Relief operations can be 'open ended'; that is, people who feel the need will come forward and be given relief. However, public works tend to take on a 'make work' character and to be socially unattractive, in part to discourage all but the 'genuinely needy' to take part. The unattractiveness of the work becomes a means of rationing. Experience of such works tends to live long in folk memory because it is socially degrading to have been forced into it.

In South Asia there is a tradition of labour-intensive public works undertaken by the landless, serfs, and poor peasants stretching back beyond the British colonial period. The fusion of these two traditions was institutionalized, originally through Famine Code practice. Raising the status of these workers has been achieved by giving them the title of 'employment guarantee' and making them available as a group of people who made employment. Make-work activities have been productively linked to asset formation. The link is provided through grafting relief operations on to a continuing decentralized, district-level, rural public works programme which has a portfolio of economic and social infrastructural investment and maintenance. The rural works programme is then a potentially powerful component of a food security programme as operations can be rapidly expanded in the face of a deteriorating situation. The existence of a strong framework of local government or administration is a necessary condition for such an effective linking of relief with public works. Availability of resources from the centre or in local budgets is also a critical element. Triggering mechanisms institutionalized as famine indicators and procedures for declaration of 'scarcity' and 'famine' are of critical importance.

Food and relief works

There is no necessary link between relief works and special food supply operations. The recent re-examination by Sen of the historical record for major famines including the great Bengal famine of 1942–4 and the Bangladesh famine of 1974 has called in to question the necessity for the traditional emphasis on food availability and additional food-supply measures (Sen 1981). However, Sen's counter-emphasis on the inability of affected households to sustain their food entitlements – that is, to feed themselves directly or finance food purchases out of income, by credit or dissaving – underscores the importance of rural works in replacing temporarily lost livelihoods. Long-established institutional practice for containing the effects of natural disaster, such as the Bengal Famine Code, have also given highest priority to additional measures, such as special rural credit programmes and tax relief, to sustain the assets and livelihoods of farmers and those whom they are likely to employ through periods of crisis. Famine Code theory, if not practice, assumes a battery of such measures to avoid or limit the disintegration of normal patterns of production and employment, and the assets on which they depend – livestock, equipment, seed, and land.

International relief agencies may be able to respond only with food, other commodities and personnel rather than money, in an emergency. These are also the resources available to support rural works which typically involve largely 'local costs' items including wage goods, predominantly food, tools, and materials. Quite recently there has been a shift from cash wages to Food-for-Work in relief works and longer-term rural employment programmes.

The special relief measures in Maharashtra in 1974 and the increased scale of Food-for-Work operations in Bangladesh since the late 1970s are widely cited recent examples of public works providing an effective response to an emergency situation. In each case an enhanced level of public works was determined upon in response to a rapidly deepening crisis, in preference to the provision of food entitlements only through direct feeding or a rural ration mechanism. These experiences also confirm the finding of many earlier studies that such emergency measures typically lead to the institutionalization of public works on an enhanced scale. It should be anticipated that relief works are likely to become, in due course, a development programme.

As the South Asian programmes are widely considered the most successful, it is perhaps appropriate to point to some of the problematic or controversial aspects of public works and in particular Food-for-Work with illustrations from these programmes.

The pattern of participation

The large Indian and Bangladeshi programmes show that public works can have a significant impact on both employment and incomes in a relief situation and as long-term rural development projects. The Maharashtra Employment Guarantee Scheme (EGS), was found to provide 10 per cent of employment amongst the poorer sections of rural society. Participants worked an average of 160 days per year, but this still left 90 per cent of households below a widely accepted poverty line for rural India. The EGS and Bangladeshi programmes have had a differentiated impact in terms of the types of poor households which participate. A widely expressed concern about public works is that the weaker and disabled are disadvantaged when entitlement is based on participation in work. However, the relatively less attractive nature of the manual tasks involved and low work norms in public works, appear to draw in those who are disadvantaged in seeking other employment, for lack of skill or physical stamina, or by class, caste, and gender discrimination. An example is the anticipated high level of female involvement in the EGS and Food-for-Work in Bangladesh.

Problems of project creation

Assessments of programmes show another characteristic tendency and limitation of the public works record. Although poverty and food insecurity are highly regional in incidence, programmes tend to benefit the relatively advantaged, or less disadvantaged, regions. Many economic, technical, and political factors work to the disadvantage of relatively backward areas. For example, the types of public works which can be carried out on a labour-intensive basis may not provide the infrastructure most needed in the area.

Public works provide a way of supplementing income from the agricultural and rural sources. However, agricultural employment incomes of small farmers and the landless are highly seasonal. For example, in Bangladesh rural works have traditionally been organized in the dry season in which employment opportunities were relatively limited. However, the period of peak vulnerability of low-income households throughout Bengal is the pre-harvest period, towards the end of the rainy season, when physical conditions severely hamper widespread public works.

Different types of works activity also have a varied employment potential. For example, South Asian experience is that the proportion of unskilled labour in total project costs is likely to be highest in the case of directly productive investment projects such as wells, re-excavated tanks and irrigation channels, reafforestation, and soil conservation. But there is a severe problem in maintaining a pipeline of fresh projects in any region, and the share of road construction, maintenance, and rehabilitation is likely to rise in consequence. The organizing of income and food entitlement for poor people around public works is constrained by the specific physical and technical factors in each local environment. Rural works as an indefinite source of supplementary employment is likely to face rising marginal costs.

Leakages

In distribution of benefits, there has been extensive reporting of misappropriation and resources diverted from public works projects. Assessments of projects suggest a number of plausible hypotheses regarding this problem. First, the problems of management for larger programmes are likely to make the proportion of leakages greater than for small, closely supervised schemes such as those operated by some NGOs. However, the scale of works alone is not a deciding factor because it is precisely the small, decentralized schemes within a large programme that are hard to monitor.

Second, programmes which have an apparently better record in directing resources to intended uses and beneficiaries have robust local-level institutions with effective participation by beneficiaries in project planning and management. In some cases where high and very localized leakages have been reported, the intended beneficiaries have been exceptionally disadvantaged groups such as 'scheduled tribes'. Decentralization *per se* does not provide an answer to problems of more effective management where councils or special bodies do not represent the interests of the envisaged beneficiaries. Some direct participation by disadvantaged groups is required.

Nevertheless, while rural works are obviously a successful component of a short-run welfare food security net, they are also likely to provide a *quid pro quo* for the politically and economically powerful groups within rural society. They expect to appropriate the greater share of the benefit from assets created and generally do so.

The advantages and problems of a food wage

Rural works are synonymous in much recent writing and the public mind with Food-for-Work supported by food aid. Food aid is used to support public works in three different ways. First, local currency proceeds from sales of foreign suppliers can be used to finance part or all of local costs of public works including wages, not necessarily paid in kind. The Pakistani (now including the Bangladeshi) rural works programme up to 1970 was organized on this basis. Second, food is supplied to a public distribution system in a recipient country, and food, but not necessarily the same commodities, is drawn from public stocks to provide payments in kind for Food-for-Work (FFW) projects. An example is the Bangladeshi FFW. Most of the wheat in the public system is imported, and food aid accounts for the greater part of imports. Third, commodities are provided for use as payment-in-kind to workers on public and also NGO-organized labour-intensive works. The even larger national FFW programme in India, comprising the National Rural Employment Programme of state-level FFW programmes, has been supplied from domestic stocks.

What are the advantages and disadvantages of a food wage? In circumstances of scarcity, payments in kind have the obvious advantage of assuring food to the households of beneficiaries, while simultaneously augmenting local food supplies. The recent experience of Bangladesh, where leakages from FFW appear to have been very large, shows that

FFW has nevertheless played an important buffering role in the rural areas. A second advantage attributed to payments in kind is that they are less likely to be diverted or delayed than cash, but the South Asian record provides little evidence to refute this view.

A disadvantage of payments in kind, where these represent, seasonally or through the year, a large proportion of the real income of households, is that participants in schemes are obliged to resell commodities to meet other requirements. This imposes heavy transaction costs on recipients, reducing the real value of the wages. Recent evaluation of the Bangladesh FFW shows that resale does occur on a significant scale where wages are paid in one commodity, wheat. The related disadvantage that the wage good may be inappropriate has been less important in South Asia, where wheat is widely accepted even where it is an 'inferior' staple.

An issue of wide concern is the potential effect on local agriculture. Injecting into an area additional food which substitutes for purchases of locally grown produce could depress local prices. There is little evidence of direct localized disincentive effects from evaluations in South Asia. This may be because markets are well articulated and therefore the localized effects would be temporary. Again, increasing the flow of grain into the area during a period of scarcity could have a positive effect. It benefits everyone who has to buy food. Significantly, reports of localized disincentive effects appear to be associated more with international emergency relief that was based either on a wrong diagnosis of local circumstances or arrived too late. It would be unwise, therefore, to draw more general conclusions on the disincentive issue from South Asian experience.

The handling problems and costs associated with organizing supply lines of food shipments from overseas to interior areas, or landlocked countries, are considerable. Delays, deterioration, and losses can lead to disenchantment of the labour forces and those supporting the programmes. Another characteristic feature of South Asian programmes is that Food-for-Work draws on public storage and distribution systems. These programmes may even have played a significant part in strengthening the infrastructure of the food system in rural areas.

Rural works and Food-for-Work in particular are subjects of much controversy. Critics focus on evidence of substantial leakages and, therefore, reduced positive impact on intended beneficiaries, particularly in large public programmes. Those who take a more positive view emphasize the need to view pragmatically particular interventions in their wider context as buffering the rural food system. Critics are concerned that in the long-run the unequal distribution of benefits from assets will exacerbate poverty and problems of food insecurity. This, however, is a criticism that could be equally applied to virtually all interventions to promote agricultural and rural development in South Asia.

Supplementary feeding interventions[9]

The greatest international effort to combat malnutrition amongst vulnerable groups in developing countries is through the direct distribution of food and on-site feeding programmes. Such interventions are distinguished by having a nutritional objective foremost in project planning. All such programmes were envisaged as supplementing the food intake of targeted beneficiaries, normally one of three categories: pregnant and nursing mothers, pre-school children, and school-age children.

The rationale for directing programmes towards children and expectant and nursing mothers is that these are among the most *vulnerable groups* within the population. More recently the objectives of programmes have received further refinement of human-resource development for human-capital investment in terms of the economic benefits that flow from improved nutritional status, particularly of children. There is also a variety of hoped-for non-nutritional benefits, including improved school attendance and performance, improved receptivity to health and family-planning messages, to name but two.

The identification of target groups as vulnerable implies an implicit food security objective from the outset. Many of these interventions, especially those directed at mothers and small children, also have their origins in responses to earlier emergency situations. These programmes were initiated by local and international NGOs in a variety of institutional contexts. Taken together, they represent a rather loosely co-ordinated patchwork. This contrasts with rural works, which are very similar throughout South Asia and have their origins in the Famine Code and District Board works programme. Nor have these interventions been closely considered for their contribution to food security. The following summary of findings is, therefore, to be regarded as tentative, based largely on inferences from evaluation studies that had other objectives.

Two general findings are, first, survey of experience with vulnerable group programmes in South Asia does suggest that they have come to be a significant component of the food security net; second, the now considerable evaluative literature suggests a broadly similar pattern of strengths and weaknesses to those of rural works, the other major targeted intervention.

Bangladesh and Pakistan now have large nation-wide institutionalized Vulnerable-Group Feeding programmes (VGF). In a similar way to FFW, large quantities of grain are targeted on eligible households in rural areas. The VGF has come to be used in a very similiar way to the FFW in Bangladesh during 1984 as an explicit countercyclical food security mechanism by increasing considerably the coverage and period of operation, particularly in districts affected by floods. Again, the VGF provides a 'leaky' mechanism for buffering the food system and augmenting the food entitlements of vulnerable groups.

More widely, since the level of operation of NGO programmes is also sensitive to increasing stress on vulnerable participants, and in many cases these have been deliberately established in backward areas, such programmes make a vital but localized contribution to food security. So far, there has been no attempt to review systematically the geographical and group coverage of all such programmes in order to attempt an assessment of their impact. A large proportion of voluntary-agency resources are additional to those that would otherwise be available to government from its own or donor resources, and are sensitive to changing circumstances. These must therefore constitute a marginal but strongly countercyclical element in the overall food security picture.

This view of nutritional interventions as a food security measure is not one reflected in the great mass of literature on vulnerable-group interventions. It is, however, one that is supported by the now substantial body of assessment evidence. This research has shown that there is considerable transfer of supplementary food to other members within targeted households. Where there is 'on site feeding' of children in MCH projects or in schools, families at least partially substitute supplementary feeding for food that would have otherwise been provided within the household. Vulnerable-group programmes therefore represent a form of income transfer in food to households that include targeted individuals. Such interventions supplement other food-entitlement programmes where these exist and have been relatively effective, for example in Kerala and Sri Lanka. Elsewhere they represent a lower-cost and possibly more effective (because better targeted) form of intervention than ration-type entitlements.

Tensions between food security and rural development

Running through all discussion of experience in South Asia there is an awareness of the problem of balancing the short term and the long term. There is seen to be a tension between short-term benefits and negative effects of various interventions and between the short term and the long run. Longer-run concerns focus on three issues:

1. The trade-off between immediate benefits from all forms of food-entitlement intervention and long-run agricultural development.
2. The trade-off between short-run benefits to vulnerable groups and the maldistribution of long-run benefits, particularly from rural works.
3. Problems of dependence of vulnerable groups and indeed countries on welfare and aid flows that cannot be sustained indefinitely.

The whole range of food-entitlement interventions involves partial subsidization or free transfer of food to beneficiary groups. These interventions have been at various times heavily dependent on food imports and food aid. If these interventions provide at least the crudest elements of a food security net for the poor, is this at the expense of agricultural development that would reduce problems of rural poverty? The more fully documented Indian experience illustrates the complexity of the issue.[10] In the 1950s and 1960s the complicated interventions partitioning food markets, rural works, and supplementary feeding prevented recurrence of widespread famine conditions. Econometric studies suggest that food consumption in aggregate was increased by perhaps 0.5 to 0.8 tonnes for every tonne of imports, whereas production of grain may have been reduced by between 0.1 and 0.3 tonnes for every tonne of food imports. Food-crisis and high-cost wage goods are factors unfavourable to economic growth. When economic linkages are taken into account the overall impact of policies appears to have been higher rates of economic and agricultural-sector growth than otherwise. More recently, India has moved towards policies of self-reliance in food through incentive pricing, exploiting the benefits from earlier investments in irrigation, and improved agricultural technology. Agricultural growth is constrained now by lack of effective demand since agricultural growth has made no significant impact on massive structural poverty. Large public investments in stocks divert resources from other areas. The massive, cash-wage Employment Guarantee Schemes (EGS) and Food-for-Work programmes are intended to provide supplementary employment *and* increase food consumption. This is because a high propensity to consume out of additional income among low-income rural households makes rural works a mechanism for increasing the total effective demand for food. Budgetary and apparent institutional limitations on the scale of operation of such programmes are factors in the recent shift in policy to use of exports as a surplus disposal mechanism.

The tensions take a somewhat different form in Bangladesh.[11] Food subsidies through the ration system and direct distribution through Food-for-Work and Vulnerable-Group Feeding depend heavily on food aid. Management of the food system is also made much easier by the considerable flow of imported food. There are, too, the problems of structural poverty and lack of effective demand for food. A transition away from dependence on food aid and the use of food to provide welfare entitlements will run into considerable difficulties. The use of domestic food to supply ration-entitlement programmes would, in the absence of alternative domestic sources of revenue, be difficult to achieve. The limited increase in employment incomes that are associated with agricultural growth would make only modest impact on the problems of lack of effective demand and food insecurity amongst poor rural households. As yet, no alternative to sustaining these interventions with food aid has been identified. Yet continuing dependence on uncertain supplies of food aid is an unattractive option.

Sri Lanka, starting from a large deficit in cereals in the mid-1970s, followed policies which removed rice subsidies from whole segments of the population. However, there is concern that these adjustments have been far from equitable in impact. Problems of malnutrition and mortality amongst vulnerable groups appear to have increased. The lack of equity in adjustments is also possibly reflected in increasing political tensions, which raise questions about the sustainability of this course of development.[12]

Food-for-Work: longer-term implications

The preceding discussion could perhaps be summed up as suggesting that on balance rural works can play a significant role in providing food security to vulnerable groups in rural areas. Much depends on particular circumstances. There are also problems which potentially increase in the long run. How do extremely poor countries sustain interventions which are currently supported by external resources? The investments that result will create new incomes, but a disproportionate share of benefits from asset creation are inevitably acquired by those controlling land and other productive resources. Some commentators would suggest that this pattern of investment will increase problems of landlessness and the rural food insecurity problem. When part of the cost of maintaining agricultural labour is transferred from land-holders to the state, this only results in better-fed landlords. The other result is an increase in strength of the landed classes.

Open-ended vulnerable-group programmes

The proportion of poor and vulnerable within the expanding rural population of South Asia has not decreased. International agencies, governments of developing countries, and voluntary agencies are therefore confronted with a dilemma. Targeted interventions, both Food-for-Work and Vulnerable-Group Feeding, represent in effect open-ended programmes rather than projects with a finite life. Structural changes are not in prospect that will eliminate the problems of poverty and short-run food insecurity which these interventions address.

Many projects were started in a particular crisis and subsequently became institutionalized as rehabilitation and developmental projects. Many of these projects are, of their kind, amongst the better managed in the developing world. They have become part too of the complex webbing of a rudimentary food security net that stands in contrast to the situation in many Sub-Saharan African countries. There is perhaps a lesson here for countries currently in food crisis. There is a powerful humanitarian case for institutionalizing and extending vulnerable-group programmes. The challenge is to find ways of ensuring a transition from dependence on external commodity assistance.

The forms of support available in, for example, an EEC context, may be broadly characterized as capital assistance, food aid, and indirect assistance through NGOs. Is further international support required for the development of food security systems that are based on local food resources?

In food security there are two important achievements in South Asia. These states can prevent agricultural crisis leading to mass starvation and disintegration of rural social structures. India now has the capacity to do this without significant external assistance. This capacity to provide a rudimentary food security net also enables these countries to handle the potentially destabilizing effects of mass migration. Ten million Bangladeshis flooded into India in 1971 and between 3 million and 4 million Afghans since 1979 have moved to Pakistan. A reasonably robust publicly managed food system is an important building block for development. But the South Asian experience offers another sober lesson: it is possible to have food security, at least in a rudimentary form, for rural and urban people while structural poverty remains on a massive scale.

Notes and references

1. The description of food entitlement systems in South Asia draws throughout on the following studies: R. Ahmed (1979) 'Foodgrain supply, distribution and consumption within a dual price mechanism: a case study of Bangladesh', *Research Report* no. 3, Washington DC: International Food Policy Research Institute; E.J. Clay (1981b) 'Poverty, food insecurity and public policy in Bangladesh', *Bank Staff Working Paper* no. 473, Washington DC: World Bank; World Bank (1979) *Bangladesh: Food Policy Issues* Report no. 2761–BD, South Asia Programmes Department, 19 December, on Bangladesh; P.S. George (1979) 'Public distribution of food grains in Kerala – income distribution implications and effectiveness', *Research Report* no. 7, Washington DC: International Food Policy Research Institute; S.K. Kumar (1979) 'Impact of subsidised rice on food consumption and nutrition in Kerala', *Research Report* no. 5, Washington DC: International Food Policy Research Institute, on Kerala; J.D. Gavan and I.S. Chandrasekera (1979) 'The impact of public food grain distribution on food consumption and welfare in Sri Lanka', *Research Report* no. 13, Washington DC: International Food Policy Research Institute, on Sri Lanka; P.L. Scandizzo and J. Graves (1979) 'The alleviation of malnutrition: impact and cost-effectiveness of official programmes', *AGREP Division Working Paper* no. 19, Washington DC: World Bank, provides summary evidence and *IFPRI Annual Report* (1986) describes ongoing research on food subsidies in Pakistan.
2. H. Alderman and J. von Braun (1984) 'The effects of the Egyptian food ration and subsidy system on income distribution and consumption', *IFPRI Research Report* no. 45, Washington DC: International Food Policy Research Institute.
3. C.P. Timmer (1981) 'Food prices and protein calorie intake: issues and methodology', in S.M. Gilles and C.P. Timmer (eds) *Developmental Issues in Indonesia*.

4. *IFPRI Annual Report* (1986) reports on the most recent ongoing research.

5. For example: R. Chambers and H.W. Singer (1981) 'Poverty, malnutrition and food insecurity in Zambia', *Bank Staff Working Paper* no. 473, Washington DC: World Bank, August; M. Mitchell and C. Stevens (1983) 'Mauritania: the cost-effectiveness of EEC food aid', *Food Policy* vol. 8(3), August; A.M. Thomson (1983a) 'Somalia: food aid in a long-term emergency', *Food Policy* vol. 8(3), August.

6. For example: P.L. Scandizzo and J. Graves (1979) op. cit. (see note 1 above); P.L. Scandizzo and G. Swamy (1980) 'Benefits and costs of food distribution policies: the Indian case', *AGREP Division Working Paper* no. 35 (Discussion draft), Washington DC: World Bank, June.

7. J.P. Mencher (1980) 'The lessons and non-lessons of Kerala: agricultural labourers and poverty', *Economic and Political Weekly* vol. 15 (41, 42, 43), special number, October.

8. This section is based on: E.J. Clay (1986) 'Rural public works and food-for-work: a survey', *World Development*, vol. 14, no. 10/11, pp. 1237–52.

9. This section draws heavily upon recent surveys of the supplementary feeding literature by: M.A. Anderson, J.E. Austin, J.D. Wray, and J. Zeitlin (1981) *Supplementary Feeding* (prepared by Harvard Institute for International Development for USAID) Cambridge, Mass: Oelgeschlager, Gunn, & Hain; G. Beaton and H. Ghassemi (1982) 'Supplementary feeding programmes for young children in developing countries', *American Journal of Clinical Nutrition* vol. 35(4), April; A. Burgess (1982) *Evaluation of Nutrition Interventions* (an annotated bibliography and review of methodologies and results), Rome: Nutrition Programme Services, Food Policy and Nutrition Division, FAO.

10. See especially: D. Blandford and J. Plocki (1977) *Evaluating the Disincentive Effect of PL 480 Food Aid: The Indian Case Reconsidered*, New York: Department of Economics, Cornell University, July; and P.J. Isenman and H.W. Singer (1977) 'Food aid: disincentive effects and their policy implications', *Economic Development and Cultural Change* 25(2), January.

11. G.O. Nelson (1983) 'Food aid and agricultural production in Bangladesh', *IDS Bulletin* vol. 14(2), Brighton: Institute of Development Studies, University of Sussex.

12. D. Steinberg *et al.* (1982) 'Sri Lanka: the impact of PL 480 Title I food assistance', *AID Impact Evaluation* no. 39, Washington DC: US Agency for International Development, October; N. Edirisinghe (1987) 'The food stamp scheme in Sri Lanka: costs, benefit and options for modification', Research Report 58, Washington DC: International Food Policy Research Institute.

LIMITATIONS OF THE 'LESSONS FROM INDIA'[1]
BARBARA HARRISS

'Lessons'

Famines are known to have visited India regularly throughout her history and into the present century, culminating with the Great Bengal Famine of 1942–3 in which 3 million people are thought to have died. Conditions which could trigger famine still exist in India. In the early 1980s India has experienced drought conditions similar to much of Africa. But India has not experienced famine on the African scale. There have been but a few hundred deaths directly attributable to starvation and publicly acknowledged as such. Many people have wondered whether there are 'lessons' to transfer from India to Sub-Saharan Africa which might reduce the impact of drought and improve the response by ordinary rural people and by governments to a crisis in food production. The idea of transferable lessons, though an attractive one at first sight, needs to be approached with some caution.

Since Independence in 1947, the government of India and the governments of her constituent states have enacted laws and implemented measures which appear to be intended to improve the ability of ordinary people to withstand shocks to the food system. Resources for these reforms have been mobilized through the fiscal system. Economic growth and redistribution have flirted in a protracted dalliance. Fundamental changes to property rights have always been 'non-operational objectives'. The practice of policy has evolved in a gradualist way. The reforms pertinent to famine management cover three areas of the planned economy: agrarian structure, infrastructure, and the systematization of responses to emergency.

Agrarian structure

With respect to agrarian structure, the constituent states of newly independent India passed legislation early in the 1950s for land or tenurial reform with the apparent intention of securing the rights to land of those who cultivate it. In turn, this reduces rental obligations, increases the producer's control over the product of his work, and enables him or her (but it is rarely her) to have a security which renders him creditworthy to official credit agencies. Much later, from 1978, attempts were made to enhance the non-land assets of the poor through targeted credit under the Integrated Rural Development Programme.

The second major structural intervention is in fact a whole bundle of policies and measures to increase food production. These gathered momentum in the 1960s. The Indian state, centrally and provincially, has sought to service the private sector. For many years, relatively large amounts of research resources in agriculture have been devoted to R & D on food crops. The Indian Council of Agricultural Research has been responsive to the products of the international agricultural research agencies and has adapted to Indian conditions some of the major success stories in wheat, maize, and rice. The state has invested in domestic fertilizer and agrochemicals manufacturing capacity, and has provided subsidized credits for private industry to do the same. It has provided subsidized credit for private rural dealers to trade in new agricultural inputs. It has subsidized and massively expanded the provision of credit for food production both through co-operatives and through the banks (which it has nationalized to centralize control). It has established and experimented with ways of improving the efficiency of the agricultural extension service. Where the private sector has either been weakly developed or the reverse – where it has dominated and exploited producers – the state has sought to regulate it, invest in competing institutions, or replace it. From the early 1960s, the Indian state also sought to modernize grain-processing technology (stores, mills, etc.) by two means: by providing subsidized credit and subsidies on capital expenditure to the private sector; and by direct investments where the private sector is unable or unwilling to raise the capital, or in conditions where strong centralized control of food grains is necessary. From the mid-1970s, a network of large-scale, rurally located stores under state control was set up.

The third set of structural measures which are important for famine prevention concern not the production of food but its distribution. Routine public distribution of food began in reaction to the Bengal Famine and to wartime scarcity in 1943. As with measures to encourage agricultural growth, state trading in food grains gathered momentum in the mid-1960s with the setting up of legislation enabling the establishment of grain supply corporations. Both central government and the governments of the constituent states have sought to control the private grain trade through regulation, restrictions on the storage and movement of grain by private intermediaries, and by trading in competition with private merchants to provide a 'fair price' public distribution system. Very rarely and only temporarily has the government sought monopoly control of grain distribution. The state of Maharashtra's sorghum marketing system was taken over by the state government for much of the 1970s. An abortive attempt was made in 1973 to nationalize the grain trade of the whole of the north of India, but abandoned after about six months. For the most part the state intervenes in a partial way, confining its operations to those regions with most assured and voluminous market surplus. The prices at which the state has procured grain have been incentive prices for wheat, have been more or less below the open-market price for rice, and have been well below the market price for sorghum and millet. It is therefore not surprising that wheat is the most important commodity in the public distribution system. But this is not, as in Africa, because of food aid or imports, both of which have been drastically reduced since the 1960s. The Indian government is thus less vulnerable than it used to be to external political pressure using food as a lever. The Indian state subsidizes a long-term buffer stock of grain which is truly massive by international standards. It currently stands at some 23 million tonnes. This reserve is equal to all the maize and millet produced annually throughout Sub-Saharan Africa. It should see India through two consecutive years of bad drought (for generalized drought has the nasty habit of threatening production in runs of years; and every year there is drought somewhere in India). In principle the retail distribution of government grain should be accessible to all. And government grain is increasingly being used not only for captive and concentrated purposes (hospitals, prisons, police, and military) but also in the last decade for Food-for-Work: 'kind payment schemes', and in the last five years for supplementary feeding schemes for children and old people. The objective of all these measures is to improve the access of people to cheap staple food and to iron out fluctuations in supply.

The fourth major structural reform concerns labour. The right to organize labour unions is guaranteed by law. Labour-intensive rural cottage and small-scale industry have been protected and promoted since the early Five-Year Plans of the 1950s. More recently, most states have enacted minimum-wages legislation, and some have legal minimum wages which do not discriminate against women. Most states now have also made child labour and bonded labour illegal. In the 1970s many states expanded their decentralized mother and child welfare schemes (combining preventive and curative medicine, nutritional therapy and family planning) using paramedical personnel. Certain states (Maharashtra is a famous example) established employment-guarantee schemes, taking responsibility for providing and remunerating work for anyone who registers. This has spread nationally to become the

National Rural Employment Programme. All this type of legislation and intervention has as its ostensible goal the improvement of the capacity of the labour force to employ and provision itself and to withstand seasonal fluctuations in access or 'entitlement' to food

Infrastructure

The second way in which the Indian state has sought to protect itself from famine is through infrastructure. Of primary importance to increases in food production has been state investment in irrigation. This has tended to take two forms: first, direct investment in dams and canal schemes (US $1,000 million from 1950 to 1980 and the same again from 1980 to 1985); second, indirect investment through subsidized credit to private producers for the energization of water-lifting technology. Irrigation reduces the seasonality of production. It can stabilize the crop production of drought-prone and semi-arid regions. It allows the expansion of water hungry crops which are labour intensive, thereby increasing livelihoods. It enables individuals to improve their control of the production process. It could enable, and in some very rare but publicized instances has enabled, landless people to group together to invest in wells and pumpsets and become small-scale waterlords and thus improve their livelihoods. Pumpsets and electric and diesel engines have become an important part of the Indian engineering industry, thereby expanding livelihoods in the urban sector.

The Indian states have also invested massively in, and continually subsidize the provision of, rural electricity. The electrification of food production (via water lifting) has resulted in the adoption in semi-arid regions of technology based on high-yielding seed, fertilizer, and pesticides, which require assured water supplies. The state has also provided rural roads and, more recently, decentralized stores and warehouses.

Under infrastructure we ought also to mention a number of acts enabling environmental protection: the Insecticides Act (1968), Water Act (1974), Forest Conservation Act (1980), and Air Pollution Act (1981), plus the creation of a central government Department of the Environment to advise on the impact of planned development.

Response to emergency

The last major way, arguably the most crucial, in which India protects itself from famine is through a decentralized mechanism of response to emergency, as detailed in the case study of relief administration in Gujarat (p. 120).

The Famine Codes, evolved in the 1880s, comprised decentralized and compulsory guidelines to the local administration for the prediction of and pre-emptive response to famine. They included detailed local contingency plans, backed by law. They were practicable enough for one ICS officer to administer in districts with populations sometimes then exceeding 1 million. And there were institutionalized career incentives for such district officers to manage these crises well. The objective of the codes was the identification and use of a parsimonious set of local, predictive famine indicators relevant to social groups locally identified as vulnerable so as to avert the necessity of soup kitchens by the prompt provision of employment on public works. The cash wages from such employment would be varied according to market prices of grain in order to guarantee the purchasing power of vulnerable people. The Famine Codes insulated vulnerable people from bureaucratic failure to distribute food. The private grain market itself was considered sacrosanct.

In the 1970s these Famine Codes were updated and renamed 'Scarcity Manuals'. The most fundamental change arises from the post-colonial state's interventionist attitude to markets in general, and concerns the regulation of grain markets, which are now far from sacrosanct. Local administrators have not only the power but also the capacity to implement rapid emergency feeding from decentralized buffers. Famine relief is increasingly comprehensive and involves not only the provision of wages on relief works as of old but also the provision of drinking water, feeding kitchens, and gratuitous relief for economic dependants, and also the creation of more discriminatingly productive assets via relief works, and the provision of water and fodder for famine-affected animals, and credit to compete with moneylenders and pawnbrokers in order to prevent the depletion of assets and to minimize the disruption of agriculture. It is said that such measures spread the drop in consumption throughout societies affected by food emergencies and thus minimize nutritional damage.

During national food crises, the central government can ban inter-regional trade. It also keeps a centralized contingency fund to finance drought relief. Last but not least in the eyes of many, India has a relatively free press, and the state can be provoked into action through the private initiative of socially responsible media.

Thus in these three main ways India has managed to avert outright famine. And the rate of increase in the production of basic staple food since Independence has indeed outstripped the growth rate of population. This is a very substantial achievement, though it still leaves food output per capita at below levels of the late nineteenth century (a period in which there were many famines) and below the average levels of famine-affected countries in the Sahel and the Horn of Africa.

If all these measures were implemented, African societies might better protect themselves against famine. But African governments are not ignorant of these ideas and many similar interventions (in crop production, distribution, infrastructure, and early-warning information systems) have been initiated in law in African countries. So we have to think about three further problems. One concerns the discrepancy between policies and implementation. The second concerns the significance of responses to acute food crisis compared with those to chronic food shortages. The third concerns the ways in which Indian conditions are different from those in Africa.

Non-lessons from India

The description of the formalities of reforms and of development planning which I have outlined above is tantamount to a panegyric, suggesting that authorities have grounds for pride and complacency. Neither is really justified. For India may have averted large-scale famine, but she certainly has not made an impression on the problem of malnutrition. Nearly half of the entire world's malnutrition is said to be in this one country. How do these measures fail to deal with this chronic problem while being reasonably good at dealing with the acute one? An answer to this question is as relevant to African conditions, as is a list of the policies which have been associated with India's impressive gains in grain production. Our answer requires going beyond policy planning to look at its implementation.

Agrarian structures

With respect to structure, land reforms have only been implemented in a few states. Elsewhere the legislation is sabotaged or evaded by landowners who own land in excess of the legal limits. In many cases tenants have been evicted by landlords rather than gaining their rights to what they cultivate; and tenurial conditions may become more insecure. The very large landless population has not been able to gain access to land. Where land reforms have been implemented, lack of property reform means that money and commodities may still be in the hands of a small ex-feudal élite. And a successful land reform may create a pauperized petty commodity producing peasantry unable to accumulate a surplus with which to invest in new technology

The production technology that has been transferred has to date been concentrated on a few crops and a few regions – wheat particularly, in the north (west) of India. Over the last decade, under circumstances where the new technology has proved appropriate, it is now recognized that almost all classes in society have been able to adopt it even though poor producers may be the last to adopt and be compelled by debt rather than by entrepreneurial zeal to 'modernize' their agriculture. But the state's research system and its infrastructure is patchy. Arid and semi-arid regions growing water-sparing crops now suffer a double disadvantage. First, there is no widely appropriate high-yielding technology for their crops and production conditions. Second, grain prices are lowered by the existence of surpluses from regions which have experienced the green revolution and thus provide little price incentive to innovate. The provision of state credit has not ended the domination over producers of merchants and moneylenders in parts of India such as West Bengal.

Changes in production are changing the environment. Environmental protection legislation is widely observed in the breach. It is said that 53 per cent of India's top-soil, and its nutrients, is subject to serious wind and water erosion (especially in drought-prone regions) and is redeposited under conditions of accelerated siltation in dams, in valleys, and at the coast, playing havoc with flood plains. Forests are being depleted at a net rate of 1.5 per cent per year for fuel, construction materials, industrial raw materials, and to release land for permanent cultivation. The department where responsibility lies has insufficient powers to halt the degradation of the environment. Action against desertification and deforestation, against waterlogging, salinity and alkalinity is *ex post*, rather than *ex ante*, and resources are utterly incommensurate with the scale of these social and environmental processes at work.

If we turn to the distribution of grain, we find that lack of finance means that grain procurement by the government may be carried out in times of scarcity (putting pressure on already abnormally high open-market prices) not in times of glut, when prices are low. And rations get cut as a result at precisely the point in time when vulnerable people need cheap grain most. In Tamil Nadu, for instance, the good year of 1980 saw 86,000 tonnes of grain purchased by the state government. By contrast, in 1982 under conditions of a one in 100 years' drought, 604,000 tonnes were purchased. Yet the eligibility to grain of targeted households had to be reduced from 20kg per month down to 5kg at the height of the drought, when the need for food was at its maximum, only to be restored again once a normal harvest had come in. The system has until recently been accused of having a massive urban bias. This operates in two ways. First, the retail outlets for state grain have been overwhelmingly urban. Second, those who subsidize the public distribution system are those who purchase grain on the residual open markets. There prices are higher because merchants have legitimately compensated for losses made by forced sales to the state by hoisting the prices of their residual grain. The 30–40 per cent of the rural population which is landless depends materially on the open market for its grain. It is these most vulnerable people who pay for the cheap rice in cities. Of late the public distribution system has begun to expand into rural areas. But the small retail outlets are unviable and need increased state subsidies to run. The same increased need for subsidies emerges when the system is expanded to provide therapeutic and supplementary feeding and in-kind payments for those finding work on state employment schemes. In some states the subsidy of the public distribution scheme is the largest item in the government's budget.

The system does not operate independently of the state and is often characterized as a corrupt supplier of poor quality 'élite' grains, such as wheat and to a lesser extent rice, inappropriate to the needs of the poor in peninsular India. Their diets are mostly based on sorghum and millet. Official procurement prices for these 'coarse' grains are somewhat casually fixed, usually at levels considerably below open-market minima and consequently unattractive to sellers. The marketed surplus of coarse grains, sporadic in time and place, is unattractive to public purchasing institutions which operate with high fixed-cost components.

The final structural reform – legislation to protect labour through minimum wages – seems exceptionally hard to enforce. There has been little change in the subordinated status of women. Child and bonded labour is still the lot of millions. And the extreme poverty of the large proportion of the population that is assetless constrains the market for manufactured consumer goods.

Infrastructure

Turning to the public provision of infrastructure, irrigation has also been concentrated in certain favoured regions. By contrast there are regions (e.g. the Coromandel Plain on the coast of south-east India) where the state has failed to prevent private farmers from oversaturating the land with pumpsets, thereby leading to a secular decline in the water table. Time and food have been bought at the expense of ecological decline. Electricity supplies fluctuate; the system is alleged to be very corrupt. The extension of electrification to farmers' fields for agricultural use has resulted in the creation of a grid which maximizes transmission losses. In turn this means that electricity supplies become a major constraint on the capacity utilization of (urban) industries. The public provision of roads and transport is a major subsidy to rural élites. But public-sector warehouses are rarely used by farmers, even by the élite landowning class, almost as rarely used by merchants and mostly used by the government for the storage of grain and by big companies for the storage of fertilizer.

Response to emergency

With respect to the Famine Codes, despite (or because of?) the existence of decentralized administrative protocols and funds, the financial resources for emergency drought/flood/cyclone relief and rehabilitation are disbursed at the discretion of central government and are frequently alleged to be late, not based on technically sound information, and inadequate. The relative merits of aid in cash or in kind and of aid in kind either as public feeding or as Food-for-Work on the creation of village-level assets, are not fully understood and still widely debated. Furthermore, famine in India is an official event which governments are generally reluctant to declare, a label to be avoided; Gujarat may be an exception. There is no reason to assume that there have not been starvation deaths in periods of food crisis. Their number, however, has not passed the cultural, administrative, and political threshold defining 'famine'. Lastly, there may be a free press which has proved itself in alerting the electorate about food crises, but it plays a dilatory role with respect to chronic malnutrition. A free press is not of significance unless there are structures of power able to translate rhetoric into action. In many states (such as Tamil Nadu) democratically elected institutions of local government were suspended for fifteen years, which has hampered the development of democratic and participative government.

These are the kinds of processes and relations which mean that fine words on paper fail to make an impression on rural malnutrition even if they may succeed in averting famine.

Differences between Africa and India and implications for 'lessons'

'Lessons' imply that actions and resources which have proved effective in India can be transferred and/or developed in Africa. There are three obvious objections to this. One follows from the logic of our demonstration that the 'lessons' of India avert famine but not widespread malnutrition. African planners may be assumed to want to avoid both famine and malnutrition (although given the use of famine as a tool of war and the priority nature of military planning, some might want to dispute even that intention. Certainly, African politicians cannot be presumed to want to put an end to chronic malnutrition).

The second objection is that to talk of 'the African state' perpetuates the idea that the countries of Sub-Saharan Africa are not all unique in ways which are important. There are many kinds of state in Africa. Some African states are successful in preventing famine. Those which suffer famine have certain common features, including being among the poorest countries, having extreme and ecologically marginal environments, declining food production per head, being locations of conflict, having relatively large agropastoralist populations, and powerless and remote agricultural populations. 'India' and 'Africa' are often code words for stereotypes which are not at all useful in the contexts of administrative and social reforms within individual countries. Yet our discussion of general transferability of experience forces us to use these code words.

The third objection is that although there are similarities (India and Africa are large, in terms of both population and land mass; both are diverse; both India and Africa suffer serious environmental degradation; both consist of many states, although India is governed as one nation), there are so many differences that affect the capacity of individuals and governments to avert famine that India's experience, even the most appropriate elements of it, is untransferable.

In this concluding section we shall examine the implications of the differences. First, let us consider certain demographic factors. India has a greater pressure of population upon land, while growth rates are certainly faster in Africa. Many Indian districts have populations exceeding that of a number of African states. Population density has implications for all the reforms which we have itemized for India. One does not have to be a population determinist to see that it is harder, for instance, to maintain irrigation systems (which intensify, expand, and stabilize foodcrop production) where the population is sparse.

A second set of differences concerns aspects of the food economy. India has greater inequality of landownership. Land and property have been private and commercialized for longer there. India has a larger proportion and absolute number of landless people, dependent upon markets, and with low and fluctuating purchasing power. The politics of assets redistribution and of 'getting prices right' have been further up the agenda for longer in India than in Africa. Local social mechanisms for coping with scarcity have been long challenged by the forces of commercialization in both India and in Africa. In India local survival strategy took the form classically of redistribution between castes within a locality. In Africa our understanding is that (despite the existence of more or less disguised sharing within lineages and/or by gender which cushioned vulnerable groups at the point of consumption), it was the household itself which operated coping mechanisms (overplanting, intercropping, granaries). Redistribution between the basic social units – tribal groups – did not take place at times of scarcity.

It has been said that the most important contrasts between the two stereotyped food economies are (1) that Africa's problem is one of production while Asia's is one of distribution and (2) that food production in Africa is in female hands and therefore neglected alike by extension workers and technologists. Enough has been said here to query these crude but popular distinctions. India's food problem is characterized here as one of effective demand or entitlement (even within the household). This entitlement is conditioned by relations of exchange and production in which women play probably as important and as varied a role as family labour and as wage workers as they do in Africa. In Sub-Saharan Africa the marketed surplus is a much lower proportion of grain production than in India. At times of food crisis and recovery (as happened in Sahelian countries in the early and mid-1970s), the local marketed surplus all but disappears. This means in practice an irresistible pressure to replace dwindling local supplies to towns and cities by food aid. By contrast, through the appropriateness of her resource base and agrarian structure India has been able to use new techniques and technologies to increase not only total grain production but also the proportion which is marketed. She has developed the capacity to resist the food aid–food imports trap in which so many Sub-Saharan African countries are now caught. This trap (set in the interest of US surplus disposal) may seriously constrain the expansion of domestic food production. The wheat trap is a separate issue from that of the lack of a readily diffusable high-yielding technology for staple food crops in the semi-arid and equatorial parts of Sub-Saharan Africa.

The third difference concerns the nature and role of the agricultural marketing system. Grain has been widely commercialized for longer in India than in Africa. This has meant that the Indian state did not have to intervene to create marketing institutions either for the inputs or for the products of the new technology, for they already existed. It has, however, had to collaborate with the mercantile sector, and to regulate it and try to control its excesses. In Africa commercialization has been accompanied by the uneven alienation of large tracts of land for capitalist farming. The marketing of grain has often been in the hands of foreign traders (Lebanese or Indians) whom independent states have wished to displace, with few alternatives but to invest in public-sector marketing monopolies. These institutions are now held to be the bogeymen of famine-prone Africa's food supply problems. Ironically, such channels of distribution are indispensable to the response to famine, given the fact that alternative marketing channels capable of handling large and unusual transfers of food are poorly developed, if at all.

Fourth, there are macroeconomic and institutional differences. India has a much more diversified industrial base than the majority of countries in Sub-Saharan Africa. And it has certainly not been so dependent, economically and politically, upon protectionist international markets for a very few primary commodities whose terms of trade are declining. Physical infrastructure (roads, transport, stores) is better developed in India than in famine-prone countries of Africa. India's success in developing irrigated agriculture has depended on permutations and combinations of public and private investible surplus, subsidized electricity, plus local social and political institutions of water management, none of which appears to exist in Africa.

Fifth, India inherited at independence both a bureaucratic tradition and a large number of trained bureaucrats. Three aspects of state administration are worth mentioning here. Some commentators hold that bureaucratic competence is better there than in Africa not only with respect to planning and decision-making at the district level but also with respect to negotiations on international markets. At the risk of digressing, there is actually a serious shortage, combined with a brain drain, of most kinds of trained manpower (not only bureaucrats but also engineers, agriculturists, medical doctors for example). Outsiders to such societies, international technical cadres with a more or less summary notion of local economy and politics, are often therefore enlisted to key posts of bureaucratic responsibility.

The Indian state acts not only in times of threatened famine but also in normal times as though it must be seen to be more accountable to and responsible for the provisioning of its people than often appears the case to analysts of Africa. Lastly, the process of state formation appears at present to be more complete in India (Assam and Punjab notwithstanding). The Indian population has a national identity. The denial of food is not used as a military tactic not as a matter of principle but because conflict over 'the national question' is for the most part apparently a matter of history. In Africa, and typically in most famine-prone countries, the boundaries of states, their social composition, the control of political power and of public resources are the objects of armed conflicts in which food is a weapon of war.

For us, the most important transferable lesson concerns emergency response, but even here we remain in a state of honest doubt. One problem is that there are no data on the costs of administering the scarcity manuals and famine codes relative to the private and social costs of the development of other aspects of India's food system. Granted that both Africa and India are having their tax bases eroded, issues of cost-effectiveness ought to be important. In Sub-Saharan Africa one response to famine has been to set up early-warning information systems. But to compartmentalize them and to divorce them from (or to refuse to plan) decentralized, pre-emptive action (which requires decision-making power and funds) seems a case of missed opportunity. Famine is a complex concatenation process requiring locally appropriate and specific pre-emptive responses. Decentralization is the more important given the costs in Africa of spatial logistics. But the decentralization of the emergency response requires local accountability. It is clear that not all famine-affected African states have appropriate regional or district administrative and political structures.

Also, cash aid to famine victims has not proved equal to its textbook task of drawing in grain supplies to famine regions via signals of demand to the private market. Both cash and the organization of physical supplies seem necessary, complementary interventions in Africa. The state and its local agents cannot be assumed to be able everywhere to dominate, to regulate and to direct the behaviour of the private mercantile sector, a reflection of the latter's political power.

The post-independence 'African state' has been widely characterized as weak, weaker than the Indian state. The African peasant would seem to have far less political power and leverage than his (even than her) Indian counterpart. India still makes periodic political appeals to a largely rural electorate. In Africa this regular exercise in democracy is not carried out, though lack of formal operating institutions of democracy is certainly not a sufficient condition for famine.

Note

1. These lessons have been tried out on a small set of people, colleagues at Birmingham and academics familiar with food systems in both India and Africa: Peter Cutler, Jean Dreze, S. Guhan, Judith Heyer, Henock Kifle, Megan Vaughan, and Sam Wangwe, to all of whom I am grateful.

7

Livestock Policy in the Sahel: Why it Must Become More Drought Orientated
Camilla Toulmin

The Sahel has witnessed harsh periods of drought in the last fifteen years, damaging harvests and causing heavy livestock losses. Such shortfalls in rain have aggravated the underlying pressure on resources from rising human and livestock populations. Patterns of farming and the extension of cultivated area have combined with the breakdown of traditional controls on land use to create an anarchic exploitation of the limited grazing available.

Rainfall started to decline in the late 1960s, after the relatively favourable climatic period of the 1950s, and reached a particularly low level in 1972 and 1973. While there was a slight recovery in the mid to late 1970s, rainfall continued at markedly low levels throughout the decade, culminating in the disastrous years of 1983 and 1984. Many parts of the Sahel received less than 30 per cent of expected rainfall in the latter year and the River Niger, which flows through the region, was at its lowest level since records began a century ago. Such fluctuations in climate have caused heavy losses to both crops and livestock. However, the effects of drought on the livestock sector are particularly long-lasting, because of the time taken for stock numbers to recover. Both periods of intense drought have seen a sharp fall in herd numbers, losses of 30–40 per cent among national cattle herds were recorded over the period of 1972–3, while 60–70 per cent of cattle holdings are estimated to have been lost in 1983–4 (FAO 1985). Deaths among sheep and goats have been somewhat less severe, given their greater resistance to drought conditions. Losses of small stock also have a shorter-lasting effect on the pastoral economy since the period required for reproduction and reconstitution of herd numbers is shorter than for cattle or camels.

Drought and the pastoral economy

Losses among livestock during times of drought can be attributed to two main causes: first, deaths among stock due to inadequate fodder supplies and, second, panic sales of livestock by herders in distress. Herds have rarely died for lack of drinking water; instead, the drilling of deep water boreholes has worsened the vulnerability of herds in times of drought through the over-provision of water in comparison with available fodder supplies.

Periods of drought are usually marked by a dramatic fall in animal prices and a parallel fall in herders' purchasing power. With grain prices rising steeply, due to drought-induced shortages, herd-owners are forced to sell ever-increasing numbers of stock in order to purchase grain for their families.

If drought conditions persist, many of the smaller herd owners will have to sell female stock and may lose their entire livestock capital in this way. Following drought, as pasture conditions improve, animal prices move sharply upwards, given their post-drought scarcity and the many demands made upon surviving stock. Thus those producers who have lost their herds in times of drought have great difficulty in reconstituting herd capital in the subsequent period.

Table 7.1 below indicates the collapse of livestock prices and escalation of grain prices in En Nahud district, Kordofan, a pre-dominantly pastoral area of Sudan, as a result of the drought of 1984–5. The depletion of livestock in late 1985 caused prices to rise steeply.

Recent studies indicate that there has been a substantial shift in livestock ownership in the Sahel over the last ten to fifteen years, with animals passing from the hands of the traditional pastoral herd-owners to those of farmers and urban-based investors. This shift in the distribution of ownership has major implications for the efficiency of pastoral resource use, as 'investment' or commercial herds tend to be less mobile and thereby place greater pressure on local grazing. In addition, many pastoralists have become impoverished as a consequence of this transfer of ownership and can only remain in the livestock sector by hiring themselves out as contract herders. Many others have either moved to towns or settled to farm.

Measures to cope with drought

Effective policy measures for helping the livestock sector cope with drought thus need to focus not only on regulating access to the reduced

Table 7.1 Price movements for crops and livestock in Dar Hamr, En Nahud district, northern Kordofan, Sudan

Crops (Index)	1983	1984	1985	
Dura	100	221	338	
Millet	100	171	241	
Groundnuts	100	150	283	

Animals (£S) (approx.)	1982	1984	1985 June	October
Sheep	50	160	25	250
Donkeys	25	30	25	80
Goats	13	16	5	22
Camels	200	500	200	1700
Horses	125	130	120	500
Milk Cows	80	100	30	300

Source: N. Meadows *et al., Dar Hamr (En Nahud District), North Kordofan Province: Report of the Sudan Aid Rehabilitation Scheme,* London: Catholic Fund for Overseas Development, 1985.

supplies of fodder when rains fail but also on measures both to prevent the sharp deterioration in herders' purchasing power as drought conditions intensify and to enable the smaller herd owner to rebuild animal holdings once pastures recover. Such policies would include temporary support of livestock prices (through purchasing campaigns as drought sets in), help in establishing local grain and food stores in pastoral areas under the control of herder associations and credit programmes to fund the reconstitution of livestock holdings after drought. As will be seen below, policy-making since the early 1970s has been dominated by other considerations, in particular meeting the demand for increased meat and milk production to satisfy domestic and export markets. Consequently, herders were no better protected from the harsh effects of drought in the early 1980s than they had been a decade earlier, as levels of livestock loss and herder destitution demonstrate.

Livestock and policy-making in the Sahel

For government and project planners, the Sahelian livestock sector is perceived as posing a problem because of its low levels of productivity per animal, low level of offtake, and its apparent disorganization, with the mobility of herds over a wide area the key to effective utilization of patchy supplies of pasture. There has been a persistent hankering after 'modern' systems of livestock production, involving range enclosure and sedentarization. The development of viable policies in this sector has partly been hampered by poor information on actual levels of livestock production, sales, and exports, and partly by the adoption of inappropriate paradigms on which the development of this sector has been based. On the one hand, officials have held unrealistic expectations of marketed output from this sector, while on the other, they have often under-estimated the actual contribution made to the economy from the large volume of animal transactions which pass through unofficial channels. Pastoralists continue to be viewed as the major obstacle to effective development of livestock production, rather than as the intended beneficiaries of such measures. Little attention has been paid to tapping their knowledge and ability to generate valuable outputs of milk and meat for the rest of the economy from the meagre resources available. Instead, the approach has been one of trying to protect the environment from the depredations of the pastoralist.

Since the end of the 1960s, planning in the livestock sector has been dominated by policies aimed at the regional stratification of animal production within the different ecological zones of West Africa. The aim of such policies has been to impose order within a sector perceived as chaotic and to provide an overall framework within which individual livestock projects could be inserted. The emphasis of stratification policies is on raising levels of meat production for domestic and export markets by, for example, reducing dry-season weight loss and by fattening male cattle, using crop residues and locally available agro-industrial by-products, such as cotton seed. Within such a development programme, it was envisaged that the Sahelian pastoral zone would specialize in breeding and would provide young animals for sale to producers further south, who would either fatten them or use them as work-oxen. In the agricultural zone, for example, credit schemes have been set up to permit farmers to buy stock at the end of the rains for stall-feeding over four to six months before their sale as meat animals. Several intensive feed-lots have also been established to finish animals before slaughter. In the pastoral zone, while research has gone into ways of reducing mortality among calves and of improving calving rates, the main services provided to livestock-owners remain those of veterinary medicine and the development of water resources. Several pilot grazing schemes have been established, but no progress has yet been made to transfer real power to communities so that they can control land use in their local area. Additionally, money has been put into improving the marketing infrastructure, by building stockyards and abattoirs and by improving transport facilities, with the aim of reducing marketing costs.

Failure of the stratification policy either to develop output and incomes from this sector or to provide the herd-owner with greater protection from drought, can be ascribed to a number of factors, central among which is its emphasis on livestock as a source of meat. Whilst a large number of animals are sold by pastoral herd-owners, production of milk to satisfy the subsistence requirements of herder and family is of no less importance amongst the herd-owners' objectives. The technical and economic viability of many fattening schemes has also been found wanting. The structure of prices for animals of different age favours the retention of male animals in the pastoral sector until the age of 4 or 5, rather than the 1 to 2 years assumed by planners. Thus neither young animals nor feedstuffs have been available in the quantities assumed. The latter also have a high opportunity cost in terms of foregone export earnings if land is diverted from export crops to domestic meat production. Animals have often not responded with sufficient speed to intensive feeding programmes nor have meat markets provided a large enough margin for such fattening to be profitable. Peasant farmer fattening schemes have usually broken even, though these have often been based on subsidized inputs, but large-scale feedlots have consistently failed to cover their costs.

Work on establishing effective grazing associations is at an early stage. A number of pilot schemes have been set up in order to identify, on the one hand, areas of land which would represent viable pastoral units and, on the other, a workable institutional framework linking that land to a particular group. Most schemes represent a reversal in land-use legislation from the free access established at independence towards the attribution of certain land-use rights to a given community. In some areas, the aim has been to allocate water ownership to certain groups, to reduce the damage done to pastures from uncontrolled access to public boreholes. These groups are then in a position to exercise control over access to dry-season grazing. In other cases, such as the inland Niger Delta floodplain, traditional grazing areas exist and are being resurrected to form the basis for pastoral units in this zone. However, there have been many problems in moving from the identification of pastoral units to the successful management of these areas, such as the enforcement of rights of access, particularly in times of drought. A number of observers noted how schemes which seemed to operate with relative success were completely unable to cope with the acute pasture shortfall in 1983–4 and the subsequent invasion of the units' land by herds from other regions. In addition, for these grazing associations to work effectively they demand the development of strong local groups able to take decisions about levels of stocking and regulating access to grazing. Such groups are in their infancy in Sahelian states. However, countries such as Mali, Niger, and Burkina Faso are now committed, on paper at least, to the establishment of village and herder groups which would be given considerable responsibility for resource use in their own area.

Post-drought rehabilitation

Following the drought of the early 1970s, several policies and projects were aimed specifically at rehabilitation of the national herd. First, a temporary ban on the export of certain classes of stock was imposed by several governments, in an attempt to maintain essential breeding stock within the country. Second, taxes on stock were lifted in Mali, Niger, and Burkina Faso, over the period 1974–5, whilst compensatory finance to governments was provided by the European Development Fund. While the latter policy probably had a beneficial effect on pastoral welfare, the impact of the ban on livestock exports is less clear-cut. On the one hand, such a ban is likely to have depressed domestic stock prices, putting the purchasing power of livestock-owners under increased pressure while, on the other hand, the moderation in domestic prices will have benefited some stock-owners in their attempts to rebuild their cattle holdings. Third, a limited amount of credit was made available with which herders could reconstitute their holdings, either through government schemes, such as that of the Caisse Nationale du Crédit Agricole in Niger, or through a variety of NGO operations. Thus, for example, in the Sixth and Seventh Regions of Mali, a consortium of voluntary agencies has funded a programe for the re-establishment of producer co-operatives, with a credit programme as a major element. Credit has been made available not only to purchase a handful of small stock but also to enable co-operatives to make large-scale purchases of grain at low post-harvest prices which can be resold to members when grain prices rise to their pre-harvest peak. Several innovatory herd credit programmes have also been set up, such as the Habbannaae scheme, run by Oxfam-US in Niger, whereby herders have been able to borrow a cow, the offspring of which are kept by the borrower when the cow is returned.

Reducing vulnerability to drought

Little attention has been paid by policy-makers to ways of reducing the livestock sector's vulnerability to drought. A few ambitious project studies after the 1973 drought recommended the establishment of reserve grazing areas, such as classified forests or irrigated land, which would be made available to particular classes of stock in times of drought. Breeding females and young calves would receive priority fodder supplies. Similarly, a recent report by the FAO suggested the establishment of grazing reserves, of cattle trekking routes and policies for the orderly destocking of herds. Others have recommended that the export of agricultural by-products be stopped during periods of drought so that these could be made available to herders at subsidized rates, enabling essential breeding stock to survive. However, exports of these commodities continued throughout the drought of 1983–5, it being argued that an interruption of supplies through a ban on exports would lead to a future loss of markets. A limited amount of animal feed has been provided as emergency aid by donor nations, but the small quantities and high costs of its distribution preclude it having a significant impact, away from the main towns.

A future framework for the Sahelian pastoral sector must include the following components:

1. The means for a rapid destocking of herds once the extent of pasture failure in a drought year is known, in order to moderate the fall in livestock prices which accompanies panic sales of stock.
2. The granting of effective control over pastoral resources to local communities, backed up by government commitment to the enforcement of local rights, and encouragement of fodder production.
3. The establishment of rural and urban savings institutions. Besides encouraging herders (and urban investors) to invest in forms other than livestock it also provides them with an incentive to destock at the beginning of a drought (since they then do not hold their savings in the form of cash, which many are reluctant to do because cash tends to be consumed easily by the family).

8

Agricultural Technology

New technology is usually seen as the critical factor in transforming African agriculture. If only farmers could be persuaded to adopt technologies – equipment or farming techniques – transferred from less-backward continents, then their and the country's problems would vanish: so the argument goes. Technology development is indeed important, especially in a transition from low-intensity agriculture to more permanent systems. But it should not be overemphasized. Technologies develop in response to pressures, problems, and economic processes. If the right technologies have not emerged, this may be because the demands of small farmers and herders have not been adequately represented or their needs articulated politically. Institutional technology development occurs as a result of the organization of its beneficiaries and their articulated demand. Colonial agricultural research effectively served the needs of the empires for industrial raw material crops. Those needs were not completely the same as those of African peasants, even if many of the latter took to growing cotton, coffee, cocoa, and groundnuts with enthusiasm. Since independence the demands of farmers and even more so of pastoralists, have been subdued. Small farmers and pastoral herders have been poorly represented politically, and governments have hardly responded to their needs. Only in the provision of social services (health, education, and drinking water), which are clear vote winners, have governments responded to expressed need. The only exception has been where multinationals or parastatals have established or managed small farmer outgrower or irrigation schemes producing foreign-exchange crops. Only in the latter cases have small farmers benefited from technology development. Pastoralists have generally been quite out of the picture.

The problem is circular. Until governments show that they can do something to develop appropriate technology, farmers will not articulate a demand for technological research and development: it will not be part of their image of what government can do for them. This is one reason why big commercial farmers often benefit from research earlier than others: they get to know about its advantages and use their much greater organizational abilities and lobbying powers to see that investments are made.

Gradually, the scene is changing: more research stations are being set up, including some in the marginal famine-prone regions. The idea that things can be done by such stations will slowly take root in farming communities. Research must generally, however, be seen as a long-term investment with only eventual pay-offs, especially in regions of low and erratic rainfall where the green-revolution technology and approach is hardly applicable.

Technology and transition

It is a commonly held view that African agriculture is in a state of crisis. Famine is believed to be a result of that crisis. The crisis is due to the process of transition from shifting to permanent agriculture. In turn, this transition is a result of increasing population pressure. If that theory, first advanced by Boserup (1965), is true, it follows that agricultural researchers should focus on designing sustainable permanent systems of cropping, bearing in mind that green-revolution-type technical fixes will not be applicable in areas of low rainfall and great local ecological variation. Nitrogen fertilizer is hardly useful in such areas, and high-yielding seed varieties are less useful than drought- and pest-resilient ones. Requirements for permanent systems of cultivation are attention to soil dynamics and deficiencies, experimentation with rotation and different cropping patterns with the objective of fertility maintenance and enhancement, and development of seed varieties which meet the farmers' needs. Investment in land, soil, and water conservation – bunding, appropriate cultivation methods, re-forestation, water harvesting – will complement research on permanent systems of farming. In the livestock sector reallocation of clear land rights to particular groups and inducements to those groups to invest in land and stock-improvement measures would be necessary complementary policies. There are now research programmes working along these lines, although this has not been the major thrust of agricultural research in Africa.

The opposite view, articulated by many social scientists, is that to speak of transition is premature. Plenty of land is available: shifting agriculture is a rational, efficient response to an abundance of land and shortage of labour. The need from science, according to this view, is for innovations which will increase the productivity of labour not land. Mechanization is high on the agenda, particularly where there are large tracts of uninhabited or 'underused' land: only from large-scale tractorization will Africa feed its population. Peasants need to be helped by appropriate mechanization – whether tractor or animal draught – to overcome the labour constraints which keep them from producing more.

Official science can help with supportive and adaptive research on mechanization and cultivation regimes which make best use of available labour. Investment in increasing land productivity is irrelevant, for the most part, and in any case does not really represent progress. The Asian model of intensification by the organization and application of labour and capital to improve land is seen as 'involution' (Geertz 1963) with limited scope for increasing incomes and rural wealth, necessary in Asia, no doubt, but not in Africa.

This debate centres around the relative abundance of land and labour. Most statistics on the subject report vast tracts of unused land. Only in Rwanda, Burundi, Uganda, Malawi, Lesotho, and Ethiopia is more than 10 per cent of all available land cultivated. Only in Rwanda, Burundi, and Malawi is one-quarter of all land used. Surely there can be no dearth of land? Apart from the doubtfulness of such figures, it has to be asked how much of the unused reserve is cultivable? How much is used as grazing land? And finally, where is the unused land? Is it accessible and habitable, or inaccessible and disease-ridden? Most countries in Africa now have both densely populated and thinly populated areas. Organizing transmigration to occupy unused or underused land is politically sensitive, expensive, and not easily managed.

The dangers of assuming a surplus of land can be illustrated by Kenya. Out of a total area of 57 million hectares only 4 per cent is said to be cropped. Nevertheless, Diana Hunt has convincingly argued the case for land redistribution as the only feasible method of providing incomes and work to the increasing numbers of unemployed and underemployed (Hunt 1984).

Certainly it is not always easy to identify vast unused areas for resettlement. Often 'unused' in official parlance actually means used by pastoralists and others who derive incomes from forests and grasslands, through honey extraction, hunting, charcoal-burning, grass-cutting, and gum arabic harvesting. Sometimes the value of this produce may be greater than the value of crops which would be produced on the land if it were cleared.

If land is not always easily available, and if some areas (Rwanda, Burundi, the Ethiopian highland, north-eastern Ghana, the Nigerian state of Kano, etc.) are densely populated and exposed in varying degrees to land deterioration, there is a vast array of situations where there is no *great* scarcity of land, but when fallow periods may be shortening, and where demands for the land's produce of all kinds – fuel as well as crops, meat, and milk – are increasing. In other areas, peasants face no scarcity at all: there is indeed an open land frontier. Trends will be in the direction of greater scarcity and exploitation of existing accessible land, so long as people remain relatively immobile, resources for opening up great new areas are limited, and absorption of labour outside agriculture is limited.

However, this conclusion does not mean that shifting cultivation is about to disappear. Paul Richards has given us a useful perspective on this (Richards 1985: 49 ff.). Shifting cultivation can be seen not as the traditional unchanging form of agriculture in much of Africa but as a response to a set of circumstances which arose during the pre-colonial and early colonial period: the spread of rinderpest, which decimated herds and thus the main alternative source of fertility (dung); the impact of colonial trade which brought new markets, the easiest way to satisfy those markets being to cultivate new land; and the advent of colonial order which made distant lands more accessible. African farmers know how to manure the soil but employ shifting cultivation because they have surplus land and little labour. Shifting cultivation should be seen not as a problem but a resource. It is always an evolving system. One finds as much innovation and experimentation in areas with low population densities as in those with high.

Richards advocates scientific research which supports farmers' own experiments, involving farmers in problem formulation. In this way a scientist can remain agnostic on the question of whether or not there is a process of transition. Basing their work on farmers' own experiments would mean decentralizing research to respond to local ecological variations. It would also mean studying how to cope with the complex interactions of soil, climate, pest, and so on, and not proceeding by isolating variables. This is feasible given 'a flexible statistical design with sufficient accuracy for estimating treatment effects' (ibid.: 143). Clearly, for any such approach to work a good understanding of farmers' own beliefs and systems of cultivation is necessary.

This approach would reject the negative judgements on shifting agriculture of colonial agriculturists and modern ecologists, unless farmers themselves are looking to build up farming systems which do not rely on shifting or fallowing. It would argue that famine is not the result of a lack of appropriate technology development, although the inability of farmers by themselves to respond adequately to changes in rainfall or fallow periods may be a contributing factor.

Biases in crop and livestock research

The most successful programmes of crop research (plant breeding, agronomy) have been either in export crops (cotton, groundnuts, coffee, cocoa, tea, palmoil, coconuts, tobacco) or, in East and Southern Africa, maize. These programmes arose and were strengthened in response to colonial requirements, either of colonial governments seeking export revenues or to supply metropolitan industry with industrial raw materials. In East and Southern Africa it was pressure from well-organized lobbies of commercial settler farmers growing maize. Post-colonial governments have achieved some diversification, but the agricultural research systems have been weakened by the absence of strong lobbies demanding effective research. The small farmers have by and large been ill-organized and have had limited access to political power. Big farmers have often seen farming as an extension of their mercantile interests, aiming at short-term profitability rather than long-term viability. They have thus not typically demanded improvements. The exceptions have been where governments have put pressure on research systems to produce results for foreign exchange earning crops, (particularly in the 1970s and 1980s, when foreign exchange has become such a critical issue), or for food crops following the numerous political upheavals which have been associated with movements in food prices or local availability. Pressure has thus tended to come from governments not farmers. Work has thus focused on governments' problems, not people's problems.

In general, therefore, agricultural research has benefited export crop producers who are generally in relatively well-endowed regions. Resource-poor farmers and herders in remote, capital-starved regions have been neglected. The major food crops, with the sole exception of maize, where this is the staple, have been largely ignored. The root crops and sorghums and millets received almost no attention until the late 1970s. By this time urban population growth had advanced to the extent that production of surplus food to feed cities was becoming a major issue in many countries. Food riots, while not as common as in Egypt or other Middle Eastern countries with more developed urban economies, were beginning to appear. Food prices became a more salient issue leading to changes of government through coup or election. The result was more attention paid to research on food crops. The International Institute of Tropical Agriculture (IITA) at Ibadan now has an effective maize programme and has produced a good mosaic resistant cassava; the International Crops Research Institute for the Semi-Arid Tropics (ICRISAT) has funded the production of a high-yielding sorghum hybrid in the Sudan.

Undoubtedly national research systems have also made contributions on food crops in recent years: these remain as yet uncollated. It may be, however, that the advent of international research organizations, with their superior pay and working conditions, has creamed off the best talent from national research systems, leaving the latter less able to produce relevant results.

The green revolution in Asia and Mexico has provided the model of agricultural development which dominates the scientific and agricultural expert communities. Becoming modern implies adopting green-revolution technology, comprising high-yielding seed varieties (HYVs), chemical fertilizers, and pesticides, that is research centring upon the continuous modification of existing hybrids in order to sustain yields and counteract pests and moisture control. The returns to adaptive agricultural research in Asia and elsewhere have been substantial. Given the variability and marginality of many local ecologies in Africa, it is doubtful whether the same type of research will have the same returns. Green-revolution technology is most suited to conditions where farmers have the means to exert considerable control over crop growing conditions. It is most cost-effective in irrigated areas or where there is assured rainfall with good infrastructure and delivery systems. It has limited applicability in much of Sub-Saharan Africa, where poor soils and moisture control problems predominate.

The successes of the green revolution in Asia are partial: India is exporting grain, but half its population has the same or lower nutrition levels than at Independence; yields have increased but small farmer incomes have increased marginally if at all; and excessive reliance on chemicals has led to soil depletion and genetic mutation in pests and diseases. If one rejects the 'large-scale technical fix' as a guiding principle (but not the possibility of technical fixes *per se*), the only alternative is to 'follow the farmer'. This may seem an exceedingly conservative and indecisive prescription if a radical transformation is required. However, there is evidence that farmers have on occasion radically changed farming systems in response to economic stimuli. The most famous recorded example is the cocoa farmers of Ghana, who did all the things thought by experts to be difficult: they migrated to new areas, organized themselves in different ways to clear land, made long-term capital investments, and marketed an export crop. Only the last activity required a foreign presence. Furthermore, the cocoa farmers did not neglect food production. Polly Hill's account achieved fame because it pointed to activities outside the paternalistic ambit of the development industry while so much policy and so much gloom about development results from incestuous debate within the 'industry' (Hill 1963). There is still a tendency to ignore indigenous change while concentrating on questions of the extent and impact of official development schemes.

It is quite possible that 'following farmers' may, where conditions are right, in flat irrigable plains, lead to adoption of a variant of the existing green-revolution formula. We do not reject green-revolution technology; we only stress its limitations.

New directions

Where farmers are clearly leading the way in abandoning shifting cultivation or in intensifying their farming system by using more labour per unit of land or making small capital investment, it is clearly relevant for researchers to follow suit. There are many cases where farmers, or 'agro-pastoralists', are increasing the intensity of land use. In semi-arid areas water harvesting techniques are used increasingly.

In many areas farmers have built small-scale irrigation works, using small dams, diversion channels, or tapping shallow groundwater for supplementary irrigation. In dry areas a more secure supply of water allows for the cultivation of higher-value crops, especially vegetables which tend otherwise to be scarce in the local diet. However, if the water itself is a very scarce commodity, one man's irrigation may reduce drinking water supplies to others. The capital required to dig a well may be out of the reach of most farmers, so that only the relatively rich are able to benefit. Finally, the vegetables produced may not be purchased by villagers or poor townspeople but by the middle classes. Agricultural development work which is designed to improve nutrition in the countryside cannot stop at the point of production but needs to concern itself with the whole food system. Irrigation, particularly small-scale irrigation, must be a fruitful area for technology development. Other promising areas in the field of moisture conservation lie in preventing water losses – reducing evaporation from soil and water surfaces, seepage losses from canals, channels, or tanks, reducing transpiration and percolation, and of course the wide field of breeding and managing crops and livestock to maximize scarce resources.

Africa's problems are different from Asia's, or those of other regions. The notion of a 'technology gap' has to be used with caution since it can wrongly imply that the direction of technological change that is suitable to Africa's diverse conditions is known *a priori*. There is no end state of technology development; constant questioning on grounds of local suitability is essential.

Research on artificial fertilization, often promoted by FAO and other donors, has been extensive. While fertilizer use is clearly beneficial in the short term where moisture availability is assured, this use is constrained by poor infrastructure and acute shortage of foreign exchange. Therefore scarcities are common, especially for smaller farmers. In the poorer, more fragile environments, the use of artificial fertilizer also allows for the abandonment of traditional methods of fertilization – manuring, crop rotation, fallowing, and so on – indirectly contributing to soil impoverishment. Research on fertilizers has been carried out largely in the context of pure stand crops. The effects under intercropping or mixed stand conditions are less well known.

Where moisture is limited, say less than 900mm rainfall (though it is the distribution in time and space not the total amount of rain which is critical), nitrogen fertilizer is unlikely to be economic. Large volumes tend to be needed; they are expensive and are usually imported. However, small amounts of phosphate fertilizer have been found to increase millet yields quite dramatically in parts of the Sahel.

Where land is becoming scarce or fallows are being reduced markedly, there is great scope for emphasis on new methods of soil management. Once again farmers themselves will probably lead the way. Farmers in Mali have adopted quick-maturing varieties of millet and appreciated the importance of dung as a fertilizer for the millet to the extent that they were prepared to pay livestock owners to graze the stubble of their fields. However, in a drought year the dung can burn the crop: research is needed on techniques to prevent this.

The study of organic fertilizing has been grossly neglected, though it is now being seriously examined in Tanzania as part of its pursuit of national- and village-level self-reliance. Ex-President Nyerere has written:

> We urged the use of chemical fertilisers; we established a fertiliser factory – heavily dependent on imported components – and we ensured that chemical fertilisers were used in growing certain of our export crops, especially cotton and tobacco. We also adopted a World Bank-assisted maize programme, which depended upon using chemical fertilisers. For the rest we left our peasants to carry on as before. We even stopped teaching compost making in our schools, regarding this as a discredited and 'old fashioned' technique, which was irrelevant to our future.
>
> The net result has been that in many places, nothing is done at all to re-fertilise our soil after it has been used, much less to improve the fertility. Our peasants can no longer 'move on' after their plot has lost its fertility; they just get less result from their sweat and – legitimately – complain that having told them to use fertiliser, we do not make it available at a price they can afford or when they need it. And sometimes the continual use of chemical fertiliser makes the soil very acidic, and thus reduces its productivity. We now get less cotton per hectare in Sukumaland than we used to.
>
> (Nyerere 1984)

Nyerere goes on to argue that not only are chemicals expensive but also, apparently, dangerous when in large quantities. While not an advocate of organic farming, he believes that Tanzania should not depend on chemical fertilizers, but learn about and develop scientific methods of agriculture which preserve soil fertility and use no chemicals at all.

Other aspects of organic fertilization are also neglected: green manuring, the planting of cover crops and forage crops, and the development of mixed farming systems. Farm forestry too has potential for both income generation and soil improvement and protection. Organic fertilization and mixed farming should also not become dogmas, but remain among a number of avenues to be explored in appropriate situations. For example, encouraging farmers to make compost trenches in their fields may be ignored by farmers if they can more easily move their fields to more fertile land. On the other hand, compost-making where fields are permanent – often those near the house – may be viable.

The success of plant breeding in India and Mexico has led green-revolution enthusiasts to rely on transferring successes to Africa. To a great extent there has been a reliance on technology transferring from outside Sub-Saharan Africa. Of the international research stations only the International Livestock Research on Animal Diseases (ILRAD) is carrying out basic research. The West African Rain Development Association (WARDA) and ICRISAT have been preoccupied until recently with the introduction and modification of exotic genetic material in rice, sorghum, and millet. While the limited benefits produced by these organizations are perhaps disappointing, this should not be surprising since even adaptive research takes much time, effort, and continuity, and the environments for which plants are being bred are extremely heterogeneous. Even the limited benefits produced by these and other national or sub-national efforts are often not transferred to the farm level because of weak communications and extension systems.

Despite some limited successes, for example in sorghum breeding for the irrigated areas of Sudan, ICRISAT is adding greater exploration of local varieties to reliance on transfer of genetic material from India. Some applied research has found as much variability in local genetic materials as in the imported range. This is an encouraging development. Undoubtedly, plant breeders anxious for quick results will continue to work with exotic germ plasm: our argument here is not that this should cease but that a much greater proportion of their time should be spent working with local varieties.

Richards has gone further to suggest that farmers could be involved (1) in genetic conservation of local plant material which can otherwise be lost as new varieties are introduced, and (2) in local seed multiplication, at which they tend to be highly skilled (Richards 1985: 146). He also suggests that farmers may prefer a variety of planting material rather than the uniformity of the pure lines which scientists tend to produce.

Following farmers is a recipe for diversity. Many farmers will plant a great range of seed varieties of one crop to suit differences of terrain, date of planting, and to maximize the likelihood of getting a crop. The most researchers may be able to do in the circumstances is to come up with one or two more varieties or even crops which may be useful additions to or substitutes within the system. But the system is likely to remain firmly within farmers' control regardless of the extent of external inputs. Researchers should concern themselves with producing a range of seeds for particular local environments and allow farmers to choose how they use them.

The organization of research

Understanding peasant livelihood systems is definitely one prerequisite for effective technology development. Farming Systems Research (FSR) is the major school of thought which has emerged during the 1970s to achieve this. Basically, FSR involves the employment of social scientists to tell natural scientists about the constraints and opportunities experienced by farmers, and encourages scientists to look for marginal improvements to existing farming systems rather than designing perfect but inoperable systems on the research station. Such understanding is necessary if researchers are to do relevant work. However, acquiring a better understanding is not the only critical aspect of research organization and strategy. Agricultural research is by nature a medium- to long-term business. Results rarely appear overnight, and are least likely to in marginal, famine-prone regions. The other critical issue is how to organize and sustain research in such regions on a long-term basis.

The recent trend has been to set up large, expensive, well-equipped research organizations, staffed mostly by expatriate scientists, with a high proportion of bills being met by donor agencies. In western Sudan four research stations are being built with World Bank loans and operated with assistance from USAID, which, in Kordofan, is paying 70 per cent of salaries in addition to other operating costs. The dangers are, first, that the research establishment will never pay its way, and that the payment of the loans with which it was built will be burdensome, second, that knowledge is accumulated by expatriate scientists working on short-term contracts, without either a clearly defined strategy of research or clearly assigned national counterparts (trained scientists do not like to be mere counterparts); and third, that without aid such organizations would collapse, or at least suffer severe dislocation if one agency withdraws support before another steps in. It is considered by the World Bank that large, well-equipped laboratories and research stations are needed to attract national scientists to poor regions. Interaction with expatriates and the lure of foreign training are other attractions. Whatever their problems, such organizations at least get a badly needed process of research going, and contribute to the development of scientific cadres with professional interests in famine-struck areas.

Given the substantial drawbacks, however, alternative forms for research must be sought. By all means provide well-equipped laboratories and stations but perhaps on a smaller scale and with priorities clearly identified. Small units could be located in different agro-ecological zones and could be added to as required. Employment of expatriates should be minimized, and the unworkable concept of counterpart scientist should be avoided, though young national scientists could be trained on such programmes rather than going abroad to research irrelevant topics for their masters and doctoral degrees. If expatriates are employed, it should only be to fill a gap which nationals cannot fill. Only thus will a long-term process of institution-building occur.

Foreign training is of dubious value unless it is at an international institute committed to research for peasant agriculture. So many scientists are trained on irrelevant, microscopically detailed topics at master and doctoral level, which is of little use to them when they return home. They need to identify major farming or herding problems and work on them from a variety of perspectives, preferably together with colleagues with slightly different disciplinary backgrounds. The major alternative to foreign training is much more local training on the job and by national universities which should use their trainees to undertake research of use to particular communities. A more comprehensive approach would no doubt seek to reform science education as a whole. We cannot go too far into this debate here, however.

Both approaches imply that scientists must acquire some understanding of what is needed by farmers or herders. In the FSR approach, social and other scientists work with farmers, through surveys, participant observation, and farm-based experimental work to identify farming systems, problems and to get feedback on the usefulness of the technical changes introduced. This process runs alongside basic on-station research. It is a peculiarly American approach to research; European research tends to have been more discipline– and station–focused. Since most African research organizations and traditions were moulded by European scientists, FSR represents a major innovation in philosophy and practice. It must constitute a welcome shaking-up of the professions involved. However, FSR in its full-blown form is expensive. It is only needed in full where scientists are not already familiar, because of social origin or education, with the circumstances for which they are producing innovations. Thus it is most needed where large numbers of short-contract expatriates are employed. As an institution becomes better established and develops good links with farmers and herders through a variety of channels, structured patterns of interaction with farmers are less needed, apart from on farm research trials. In between is a range of situations in which aspects of FSR can be used without accepting the whole concept, or the expense of operating an FSR system. Other stimuli, the political system, the ebb and flow of particular farming problems, will also contribute to identifying research topics. FSR is a complicated and formal way of persuading scientists to listen to the demands of their environments; when there are other cheaper and less formal ways, they should be taken. Above all, an active peasant political lobby will act as the best insurance that researchers are concerned to deliver appropriate innovations.

Mechanization

Mechanization constitutes the other major plank of agricultural development strategy. Part of the argument for mechanization, particularly tractor mechanization, derives not from an analysis of labour constraints but from the more general inability of African states' food production to keep pace with population growth and demand. Janos Hrabovsky writes:

> Part of the problem stems from the rapidly rising ratio between food consumers and producers through the process of urbanisation at a rate of about 5 per cent per annum as compared to a rural population growth of 2 per cent for the period 1975–85. If you add to this the fact that in recent years each percent growth in rural population brought forth only 0.5 per cent increase in arable land, and that yields were close to stagnant, then it is clear that a total output growth of 1.8 – 2.0 per cent leaves a big hole in the food balances after a period of ten years, when demand was rising at around 3 per cent or more per annum. It is this consideration which forces the issue that small improvements in traditional systems of farming and the insistence on soft technologies will just simply not fit the bill.
>
> (J. Hrabovsky, personal communication)

Through bitter experience it has been learnt that tractor mechanization is a heavily constrained option. It is usually more expensive than the tractor salesman projects. Machinery has a very short life; maintenance is difficult (facilities and skilled artisans are absent, spare parts constrained by foreign exchange); fuel supplies may be problematic, especially in remote areas; and soils may be inappropriate. Tractors and combine-harvesters often have to be heavily subsidized; they are also scarce, and as a result benefit only the few, usually an urban-based farming élite with privileged access to soft credit and import licences. Tractor-hiring services for small farmers are always said to be exceedingly difficult to manage. Group ownership may hold out better prospects, as in northern Ghana where the Agricultural Development Bank was prepared to lend to small groups for the purchase of a tractor and even a combine-harvester in the rice farming boom of the 1970s. The problems of such co-operative schemes are funding the replacement of worn-out machinery and the organization of maintenance. If the management problems are solved, there is no reason why such schemes should not work, as in the case of the Qala en Hanal refugee settlement in eastern Sudan. Here Euro-Action Accord manages the tractor-hire service, with participation from the beneficiary villages. Once again we have an indication that it is management or organization, not technology, which is the scarce factor in agricultural production. On the other hand, where conditions are technically suitable, as in the Sudanese clay plains, at least in those areas which are accessible, the tractor can make a massive contribution to national food supplies. But famine can of course still occur where a national surplus is regularly produced through mechanization – witness Sudan in 1984–5. With declining oil prices and over-production of machinery in western industrial countries, the tractor will become more attractive in coming years. It is all the more important that the management problems of tractor-hire schemes be tackled with urgency.

Advocates of equity tend to favour animal draught power on the grounds that it is not capable of creating inequality between producers. Animal traction has only taken firm root in North Africa, Southern Africa and Ethiopia. Elsewhere there have been many foreign-funded attempts to promote it during this century, but few have long-lasting effects owing to problems with the availability of draft animals and their maintenance, particularly during times of drought. There is a need to incorporate a concern for animal traction into national experimental programmes run by national organizations. Thus far, national programmes tend to be more interested in the tractor. Professionals and politicians have been unwilling to take animal traction seriously, believing it to be second-rate technology. Once again it is only if peasants are organized to articulate a demand for such technology development that it will take its place alongside tractor mechanization programmes.

Where land is available to peasants, animal traction may enable them to expand and produce a greater surplus, which in turn enables more to be stored for lean years as well as more to be sold. However, cultivating more land depends on overcoming the critical labour bottlenecks. If draught power (animal or tractor) does not contribute to this, it will not enable expansion. It may still, on the other hand, improve cultivation practices such that yields increase or stabilize. The financial advantage of animal traction is not always as obvious as advocates claim. R & D must be directed at alleviating the bottlenecks or releasing nutrients in the soil if mechanization is to be economic. The clearest case for mechanization is where heavy or crusted soils could not otherwise be cultivated.

We can conclude from this discussion of mechanization that it will contribute to famine avoidance only if surpluses produced can be stored nd distributed in times of need; in this case the contribution will be indirect. A direct contribution to the economies of famine-prone households and regions can be made if there is effective management of tractor hire or group ownership by peasant farmers, and if animal traction is specifically dedicated to the alleviation of labour bottlenecks (e.g. weeding or planting) within particular farming systems. Even here economic relationships and prices must allow for effective storage of part of the additional crop produced if greater food security is to be achieved.

Irrigation and water management

Advances in green-revolution technology elsewhere in the tropics and subtropics have depended critically on irrigation, just as agricultural development policy has focused almost exclusively on large-scale irrigation schemes, following on the success of the Gezira Scheme in Sudan, and in view of the availability of large, untapped river systems elsewhere in Africa. There has been much criticism of large-scale irrigation and its bureaucratic management during the last decade, which has ignored its vital (or potentially vital) role in food security. Critics of large-scale irrigation tend to advocate support for existing traditional irrigation systems (e.g. the *fadama* or river bottom land in Nigeria, the small-scale river-bed cultivations on many rivers), and the construction of small dams and tanks. Timberlake, for example, points out that South African agriculture had half a million small dams, and attributes its success partly to this fact (Timberlake 1985: 85–6). However, these dams are mostly located on substantial private commercial farms. There is no evidence to suggest that communal management of a small dam, which would be necessary in most smallholder areas in Sub-Saharan Africa, is intrinsically easier than the management of a large-scale scheme.

Where large-scale schemes would wipe out existing well-adapted agricultural systems, they should be approached with extreme caution. The costs are almost always more; the life of the dam almost always shorter, and the returns almost always less than estimated prior to construction. However, large-scale investments should not be ruled out. A possibility which has been little explored, with considerable consequences for food security, is the switching of large-scale irrigation schemes to 'irrigated-dry' production in drought years, as is now policy in India. This implies shifts in cropping from crops demanding much water (e.g. sugarcane or rice) to less-demanding ones – coarse grains or oilseeds for example – to make the most of water available across the widest areas and to substitute essential food crops for luxury or export crops in times of need. The point we wish to emphasize here is not that large-scale irrigation is unviable or undesirable; on the contrary, it has great potential for enhancing food security. But attention of engineers and politicians has been too much focused on the large scale, too little on smaller sources of irrigation water, tanks, boreholes, and even less on what now comes under the rubric of 'water harvesting'. Controlling run-off to reduce the destruction of topsoil and conserve water for supplementary irrigation, and collecting rainwater to use for both domestic and farming purposes, is seen by many as an essential weapon in the fight against hunger and the degradation of land. Three general conclusions can be drawn from a recent book about the subject (Pacey and Cullis 1986).

First, harvesting of rainwater from roofs and slopes has generally been a last resort as a source of water for domestic or farm use. Thus the best-developed systems are where there is practically no other source (e.g. for domestic use in Gibraltar and Caribbean islands). In arid zones where terracing and bunding has traditionally been crucial for cultivation (e.g. North Yemen), farming is quickly abandoned when a better-paying and less-onerous source of income is found (like working in the Gulf). In the case of domestic water supply the diseconomies of scale in small tanks, the doubts of householders and local authorities about the cleanliness of the water stored (an important area for extension education), and the fears of spreading mosquitoes and disease, have often put a brake on public and private investment in domestic water tanks.

Second, though rainfall collection is generally regarded as an inferior option both for domestic supplies and farming (i.e. there is a switch to piped water when possible), there are many areas – particularly remote arid and semi-arid areas – where existing alternatives are even less preferable: women spend a great deal of time fetching water over long distances and crop farming carries on despite low yields and soil loss. Further, many of these areas (e.g. much of the Sahel) have non-sandy soils with a high run-off and considerable erosion as a result. In these areas, water harvesting is essential for preventing desertification, for afforestation, and also serves as an insurance for farmers. In very good and extremely bad rainfall years, crop yields on land where run-off farming is practised are no different from the surrounding land; but in a medium-to-bad year they are often substantially better.

In these areas, therefore, rainfall harvesting clearly has a major role. It is often already practised and the challenge is to develop acceptable, effective, and low-cost improvements. In run-off farming the problem is the low, short-run private return to the (often considerable) amount of labour required in moving earth and rocks to create the bunds, contour ridges, hoops (semicircular catchments on slopes), and so on. This is especially the case where household labour has alternative employment during the slack season (e.g. through migration), or where the earthworks are meant primarily for soil conservation or afforestation.

Conclusion

What conclusions can be drawn from this chapter on technology development? We are arguing, first, that technologies for food production do not receive adequate attention from the institutions of agricultural development. This is likely to be overcome, however, as the problems of producing a surplus to feed the cities become more acute. Research will inevitably concentrate on the more fertile and less marginal agricultural areas where the greatest surpluses can be obtained. There is thus a major problem in establishing and, more so, in sustaining research in marginal and famine-prone areas. Governments and aid agencies can help by making long-term commitments to such research programmes. They are necessary for survival, since the inhabitants of such areas cannot be absorbed in either the more fertile areas or the towns without much greater cost and social and economic dislocation.

9

Famine and the Control of Assets

In this chapter we wish to consider a somewhat neglected dimension of development strategy, namely change in the rules and regulations of access to and use of productive resources. The importance of land, water, trees, and grazing to the resilience and well-being of rural households has been a recurring theme in previous chapters. How they find employment and whether incomes can be maintained in times of need is another.

In discussions of the economics of famine, as by Sen (1981), the changing positions of rural households in the face of famine is discussed in terms of 'entitlements'. The idea expressed is that the possession of money carries an entitlement to the purchase of goods, while other assets carry an entitlement to exchange or sale. What goes wrong in a famine is not that there is an absolute scarcity of necessary goods or money but that they lose their entitlement value. From this observation, clear policy implications follow, as we have shown.

Rules of social entitlement

'Entitlement' in everyday usage is something rather different from that in sociological analysis. An entitlement to a piece of land or to the fruit of a tree, or to the services of another person, is the result of a legal or customary status or contract. To be meaningful, an entitlement must be enforceable. A person must be able to exercise his or her rights against other people in respect of their entitlement. Entitlements may be matched by obligations, which are equally enforceable. To give an illustration, in Botswana a married man is entitled to a piece of arable land – if he can find one – which will be allocated to him by the land-allocating authority. He is obliged to use it, and if he fails to do so for five years it may be allocated to someone else. This is the law of the land, enforceable by taking action through the courts. Obligations are better illustrated by another example from the same country. A widow is entitled to have her land ploughed for her by her sons or late husband's brothers who have this as an obligation upon them and may be taken to court if they fail to oblige. Entitlements are part of the social norms of a society and can be examined as such.

The notion of entitlement in this sense clearly covers many of the same phenomena discussed by Sen but extends both analysis and prescription in important ways. Social entitlements are about rule-bound behaviour. People can only be said to have an entitlement to work if there are rules to enforce the entitlement. India's Employment Guarantee Scheme, under which the central state is obliged to provide work if groups of people come forward and ask for it, is a rare modern example of such an entitlement. According to this definition of entitlement, commodity exchange on market principles is what happens when no entitlements exist.

If we look at the entitlements and obligations that have been built up over time in any society – again Botswana can be taken as an example – many of them will be seen to deal with the risks, uncertainties and hazards which people face in their particular environment. These will not in any sense be separate from a whole range of entitlements and obligations about family survival, mutual defence, or other social concerns which cannot be neglected in any proposed changes. But the means whereby people attempt to make themselves less vulnerable can be the starting point for analysis and prescriptions.

In the arid lands of the eastern Kalahari and the Limpopo catchment area, no traveller may be refused a drink of water for themselves or their beasts; only extreme hunger in the homestead is an excuse for not providing a meal. Travelling in search of lost beasts or fresh pasture is so common an experience that these rules can be seen as a form of reciprocation. The obligation to feed the destitute if they turn up at one's gate might be said to reflect a recognition of the element of luck in the fortune of any household. Rules of this kind moderate the effects of a harsh environment and add to a society's resilience. They bring quite close the social ideal of an entitlement to food. However, we know from Jackson's historic study of famine in Kenya (Jackson 1976) that social entitlements of this kind can collapse in the face of extreme privation.

Equally important to the resilience of an economy are the rules governing productive resources, like those which provide long-term security to producers, by spreading risks, sanctions against bad behaviour by neighbours, and protection of essential supplies and other contingencies. It is not just equity of access which provides for security, though this has been the primary concern of social reformers throughout the world. In complex environments equally complex sets of entitlements and obligations build up, some of which can be seen to serve these 'resilience' ends.

The customary rules surrounding livestock in Botswana illustrate these points. Long-term security is sought in rules of common access to grazing, an entitlement vital to small herd owners who cannot afford to purchase private ranches. Risks – partly resulting from competitive use of common grazing but partly the inherent risks of disease and localized rainfall – can be met by a herd-owner by having some of his beasts herded by another person under the *mafisa* arrangement. The other person is entitled to use the beasts and to the first calf born while a cow is in his care. So this arrangement serves also as an important means of access to livestock.

Bad behaviour by neighbours, such as cattle rustling or trespass into cropland, has been countered in Tswana society by the development of formal legal institutions with established fines and penalties (featuring also an elaborate descriptive language for cattle recognition). The development of abattoirs and foreign markets, which greatly facilitated the disappearance of stolen animals, means that branding or ear-marking have in recent years had to be added to the identification system to make it work beyond the sphere of local knowledge.

Reserves of grazing or of thatching grass have in the past been set aside for seasonal or special use and protected from trespass (O'Dell 1982). It is not clear how this protection was achieved in Botswana, but in Lesotho – where there is a similar language and culture – the same problem was solved by putting the reserves into the charge of one village elder who was motivated and rewarded by allowing him to keep a part of any fines exacted in the prosecution of trespassers (Feachem *et al.* 1978: 223).

These are but some of the Tswana customary and legal entitlements and obligations that can be seen to have a bearing upon the ability of communities to build up resilient economic activities in an arid environment. Numerous examples from other places and cultures could have been cited to illustrate the same points. The Tuareg, a North African cattle people discussed by Swift make similar provisions for sharing risks (Swift 1974). The complex arrangements which Richards describes for the organization of agriculture in West African environments (Richards 1985) can also be expressed in terms of rights and obligations, though he does not choose to separate out this aspect of the farming system from a wider social and economic analysis.

One other point has to be made about entitlements and obligations. Tswana society has a well-developed set of legal institutions, aspects of which were well-established before colonial rule. In this case legal sanctions support entitlements and obligations. This points to the possibility of using legal reform as a means of present-day development of new entitlements. However, in most places, including Botswana, there are social norms that are not effectively backed by a formal legal system but are maintained instead by reciprocal arrangements, negotiation, or confrontation. Sometimes these are the prevailing means of social control. Even when entitlements are backed by law they will not survive unless beneficiaries are prepared and able to activate them. This brings out another point about the possibilities of development of new entitlements. Any new norms have to be within the power of ordinary people to activate and enforce.

The examples from Botswana also show that entitlements and obligations are subject to change and sometimes effectively break down. In one or two instances the past tense was used in the description. However, even these few instances of present or past arrangements are enough to indicate the diversity of measures which have been taken by people to make their farming systems less vulnerable. No economic or social theory would have provided much idea as to how to design such measures. It is probably safe to assert that no settlement schemes or livestock development schemes designed by government or an aid agency ever start with as sophisticated a set of legal (or customary) entitlements as can be found by careful observation of almost any existing rural society.

Today, governments are likely to be the initiators of legal reforms and enacted entitlements. Measures may cover important new resources such as tractors or credit, but the old resources, such as land, water, and labour, remain the key to rural life for most people in drought-prone regions. In most countries some element of customary law is left in place and does in fact change in the process of application, although the official view may be that it is a static inheritance. New policy on entitlements is applied through statutory law and national-level legal institutions. If there are to be sensible arrangements for secure rural production and welfare, these have now to be worked out by governments. The problem is that the prevailing development ideologies do not provide much guidance as to how it should be done.

Ideological influences

The dominant development ideologies, whether leftist or rightist, have, for understandable reasons, put economic growth first and equity second – often a poor second. Vulnerability and security issues have been very much an afterthought, usually arising in response to bad experiences, as we have shown. Leftist and rightist ideologies are also self-justifying and opposed. The adoption of either discourages critical thought about the needs of particular situations or people. Rightist and leftist prescriptions on land – which often have a strong influence upon policy – seem to be growth-centred almost to the exclusion of other values.

The most stridently asserted prescription on land policy these days derives from the classical economists, who saw a natural evolution in land rights, under growing conditions of scarcity, towards free-market principles. According to this view, a free market encourages the rational deployment of all factors of production, capital, labour, and skills as well as land. (For current expositions of this view, see Currie 1981 and Feeney 1984.) A free market allows owners to take decisions that are most efficient from a national point of view because it ensures that land is put to use by those who have the resources and skills to be able to use it.

The opposite, a collective-ownership model, can be based upon a similar deployment argument. There are two elements to it. The first is that where land is under private ownership in small plots, deployment of labour and skills will not be efficient because for much of the year labour is idle. However, under collective ownership and in the absence of boundaries, off-season labour can be used for land improvement and other activities in the common good (Gray 1978). This can also be achieved on large private estates, but the collectivist argument emphasizes the motivational aspect. On a collective, all stand to gain from collective endeavours and will be prepared to make some sacrifices to achieve these gains.

The private and collective models of land tenure development are attempts to reflect human response to scarcities. They cannot be dismissed simply by reference to the special properties of a place. It cannot simply be argued that Africa is different. Nevertheless, there do seem to be a variety of practices evident on that continent which reflect the play of present-day forces and influences upon an older tradition.

If a common pattern can be found in customary rules of access to arable land in Africa, it is probably that membership of a socio-political group carries an entitlement to land. The practice is common enough to attract the misleading term 'communal land tenure' in English-language commentary (sometimes also called 'corporate', as by Cohen 1978). It is misleading because it does not usually imply that the land is worked communally. Indeed, the rights of individuals or families to cultivate on their own account is the more normal feature, but use rights take a variety of forms according to circumstances. Two examples can be cited.

The Tiv, according to Bohannan and Bohannan (1968), do, or used to, practise shifting cultivation. Access to land took the form of entitlement, not to particular pieces of land but to *a* piece of land that would change with every outward shift from the settlement, but always bearing a roughly constant relationship to the same neighbouring cultivators.

The principle of entitlement to land for members of a group was also apparent among the Amhara in pre-revolutionary Ethiopia, where, according to Hobden (1973), in crowded terrain the position of an aristocratic ruling class (or caste?) was rendered less stable and exploitative than appeared to many observers by the fact that existing landlords could have their control of land challenged by any amongst a vast range of descendants of the original settlers of the terrain who each maintained entitlements to land.

Where there are use rights there also needs to be someone with *allocative or regulative rights*. Customarily this was often the prerogative of chiefs; nowadays it is often some more specialized arm of the central state.

Did these allocative rights in the past allow their holders to sell land? Logically, if land is not scarce, there will be no opportunity for sale, and the chiefs' interest in allocating land will be that of accumulating a political following. The interests of those with use rights in belonging to a group rather than moving off elsewhere would be mutual defence and support. However, in the face of growing scarcity many land-allocating authorities have taken to exacting a price of some sort, usually in the form of 'gifts'. However, if the land goes to strangers, this is often resented by those who are properly entitled to land. A southern

Sudanese academic recently noted (in conversation), 'a wealthy or powerful stranger who, it is suspected, has obtained his coffee plantation by making a payment to his "friend" the chief may find his crop destroyed by fire.' So there may be a counter-pressure at this point too.

The present tense has been used to describe these systems because some of them remain, though as Cohen points out, in many places the entitlements involved are changing (Cohen 1978).

Three socio-economic forces can be seen underlying pressures for change. These have been discussed in other parts of the book and can be summarized, in so far as they are influences upon the direction of institutional change, as follows:

1. Population increase leading to increasing pressure upon land and other productive resources. Effects include subdivision of holdings, abandonment of fallows, curtailment of shifting cultivation, sedentarization of nomads, and marginalization of remaining nomads through the expansion of cultivation into increasingly arid areas. Each of these measures introduces new interests that have somehow to be accommodated within the processes of social control and law.
2. Increasing interest in and penetration of commodity production and market exchange (or state-controlled exchange). The user values of productive assets will change to reflect opportunities for income generation as well as consumption and local exchange. This in turn can lead to different interests emerging between those who are in a position to exploit market opportunities and those who are not.
3. A third factor is an increasing geographic scale of social interaction and physical mobility. This results in part from the requirements of commodity production, and in part from the wider boundaries of new states. This provides business and professional classes with opportunities to move into different land areas. Most people may not have the resources to move but note the case of Ghanaian cocoa farmers (documented by Hill 1963). Also mobile these days are the refugees fleeing oppressing political regimes.

The third factor, in particular, has consequences for local forms of social control. Unless deliberately reinforced – and this has not been the trend – local sanctions cease to work. The face-to-face judicial procedures which have in the past featured so prominently in many parts

of Africa depend upon the ability of people to confront their protago-
nists directly and verbally in a commonly accepted local arena. Such
procedures are ineffective in the face of strangers, be they traders,
business persons, administrators, or teachers who can run away or get
themselves transferred or appeal to distant courts and unintelligible
written legal procedures.

Patterns of change in customary rights

Actual directions of change in different countries are the result of a
mixture of the ideological prescriptions that were fashionable at the time
of reform and pragmatic response to emerging pressures. This has led
to four distinctive present-day situations.

Market, or constrained market

In colonial Uganda in the 1920s the allocative rights of the Baganda
chiefs were converted into freehold rights (Mamdani 1976). As in the
case of seventeenth-century Scotland, the aim appears to have been to
create a wealthy feudal aristocracy loyal to a foreign crown. However,
while Scottish aristocrats and their successors are still enriched by
unearned income from their tenants, subsequent legislation in Uganda
turned the peasants into tenants of the state. This is one example of
reform due to pressure from above.

In Kenya – apart, initially, from the areas of white farmer monopoly –
the Swynerton plan opened up a market in land, but on a small farm
model. This was the most conscious case of emulation of the English
freehold legal system as an exemplar of capitalist development. It was
also a conscious attempt to create a landholding class of 'yeoman'
farmers as a basis for political development in Kenya. However,
although the colonially established measures have been largely successful in their aim, the resulting market in land is in practice constrained by
rules limiting the market in any locality to people of the local language
groups.

Allocation

In other countries, among which Lesotho is perhaps an extreme
example, the principle of entitlement to land is rigorously maintained.
Land allocated to individuals reverts to the authorities upon the death of

the holder to be reallocated to those who have not yet attained their entitlement to three fields. This may be a case in which pressure from ordinary men, more than half of whom are obliged to migrate abroad as miners (Feachem *et al.* 1978), for maintaining access to land for their families, has taken precedence over pressure from the élite for privileged access to land as a productive asset. In Botswana and other countries, although allocative procedures are maintained for some designated areas, a state leasehold system has been allowed for other areas. Land in the one case may be seen as an aspect of welfare, while in the other the demands of the monied classes can be accommodated.

Leasehold from the state: a modified form of allocation

In several countries, including Nigeria, land has been nationalized, and allocation by the state to individual users is on the basis of leasehold. The state also has the right to appropriate land for public purposes. Rights established under these provisions exist alongside customarily established rights and may take precedence over them. This can put the new élites in a position of power (Francis 1984) and may in practice downgrade the notion of entitlement to land for poorer people who only gain access to state legal institutions with cost and difficulty. Those who are in a position to obtain a lease from the state are in practice doubly privileged because such title may facilitate access to loans and subsidies.

Collectivization or partial collectivization

In Tanzania, Mozambique, and Ethiopia, as already mentioned, collectivization has been attempted with varying degrees of commitment and persistence. Although the principle has been adopted, attempts to achieve collective organization of production rather than consumption have generally foundered at the experimental stage. This may have been because collective labour could not achieve significant economies over individual efforts in dry-land agriculture. It may also have been because improved welfare was in any case the primary objective.

Where from here?

Tenure systems come, adaptively, to reflect the realities of political economy – with entitlements to purchase confined to particular ethnic groups, effective priority to economic élites, protection of the interests of migrants where these are powerful, and in other ways.

If the lessons of vulnerability of people at the margin are taken seriously, the kinds of institutional change which could make a difference are those which help to secure entitlements and increase the control of these groups over common resources as well. The necessary measures need to be worked out for each situation but might include the following:

1. Ways to ensure equitable access to arable land and avoid privileged procedures.
2. Measures to provide women with legal status and title commensurate with their role in food crop production.
3. Means of creating institutional frameworks within which there can be regulation and investment in public goods such as grazing and trees.
4. Measures, in socialist systems, to encourage the willing participation of farmers on a group or individual basis, and, second, ensure that the collective can maintain adequate resources and initiatives.

These are the big issues, the issues of equity. In many ways equally important and even more neglected are the complex norms that were discussed at the start of this chapter. These 'fine tune' farming systems to their social and ecological environments. Control of the movement of cattle in arable lands, the preservation of areas of thatching grass, protection of shade trees, are examples of norms that have been enforced in different places and which might still be valued. Such measures have equal status to rules of access to arable land and should in fact be seen as complementary to them, part of an overall framework of entitlements and restrictions within which farmers work. They always require sanctions enforceable at a local level.

There may be customary norms that are now regarded as being incompatible with certain aspects of farming system development. In Botswana, for instance, there is now pressure to allow farmers to enclose their arable holdings all year round instead of reverting to commonage during the dry season – to enable moisture conservation activities to be conducted by the farmer. But the need for a complex of rules about cattle movement and land use is not questioned, and is indeed being reinforced, as we shall see.

Institutional changes to accommodate multiple goals

Where market systems of land allocation have emerged or have been established in law, they can still be constrained, in the interests of social justice, by setting limits on the allowable size of holding (as in Indian 'ceilings' legislation). Controls on conditions of rent, practised in Uganda, is another approach. In the interests of soil conservation, government can be given the right to impose cultivation regulations, order conservation works or carry out such works and make the owner pay. This is the case in America, that haven of free enterprise, and also in white South Africa. In both these cases the farmers are few and the state powerful, so this balance of entitlements and obligations may be difficult to achieve elsewhere. An alternative approach would be to increase the powers of individual farmers whose land is negatively affected to take action against offending neighbours, or to empower local-level bodies to act on behalf of the community.

There is a case for modifying allocative systems, in the interests of both social justice and efficiency, to entitle women where they are now *de facto* in charge of agricultural activities, or to provide access for those capable of utilizing land when too much is allocated to those who are under-utilizing it. Neither measure will be easily achieved because there are strong vested interests supporting the status quo in any allocative system. Reform can be approached through a review of the allocating authority and the interests that it represents.

In allocative systems as well as market systems, landholders may need to be constrained when they fail to maintain their land or when measures like contour bunding need to be taken in the common good. In both cases the entitlements of a landholder have to be counterbalanced by entitlements which neighbours or some agency of the state can exercise as a constraint over him.

Decisions are also needed for use of remaining commons. Here the need for checks and balances in entitlements between individual users and between such users and public authorities is becoming obvious. Some of the options are considered below in sections dealing with grass, trees, and other goods that are often held in common. Here we can simply consider which authorities are likely to be involved. The problem with commons is that many people may suffer the results of degradation but no individual initiative leads to sufficient improvement for the initiator, or indeed for others, to make it worth while. What is true of use or investment is also true when it comes to prosecuting people who are breaking rules of entitlement. So without special encouragement we cannot expect an individual farmer on a group ranch in Kenya or a herdsman on an improved pasture scheme in Nigeria to take action against neighbours and kinsmen who are breaking the rules.

One traditional means of providing special encouragement is, as we have seen in Lesotho, for the community to agree to the appointment of a prosecutor who gets an adequate personal reward for undertaking the hazardous business of prosecuting offenders. A users' association, if there is sufficient incentive for people to join, can also act on behalf of the common good. However, it is likely that grassroots organization and initiative will need to be backed by some state-level organization and powers.

The issue of how to re-establish controls on grazing has been extensively researched and discussed in Botswana. The present line of approach seems to be that local communities are given responsibility for operating controls. To do this a government commission would be necessary to adjudicate boundaries between communities and register entitled users. Within these boundaries the community, helped by planners, would designate areas for arable development, for pasture, for firewood lots and access routes (Government of Botswana 1981). Agreements would presumably be enforceable in local courts and would pave the way for group or public investment in such items as trees or fences. However, this attractive concept has proved difficult to implement.

Collective systems have in principle the means of making common-good investments and often do have a good record in welfare provision and environmental improvement. In this case the entitlement problems are different. Soviet and Chinese as well as incipient African experience seems to indicate a need to work constantly at improving individual rewards and entitlements. In collectives market principles in the employment and motivation of labour are in fact replaced by rules, and the problem is to build in adequate rewards, differentials and penalties to motivate people to produce the goods and common services.

Trees, water, grazing, credit, and labour

We can now consider the development of rules of control and use of other productive resources such as trees, water, grazing, and credit. Though it is in a rather different category, we also consider labour.

Trees

Although always an important resource in the eyes of most rural peoples and seen by them as having multiple uses – shade, fodder, fruits, and medicines, as well as the more commonly recognized building materials and fuel – trees have only recently come into fashion in the development industry. Now that they are established as a good thing for the environment and for the economy, questions about how they should be grown, looked after, and used need to be addressed. Custom can sometimes be a guide since a great variety of rights and entitlements can be found in different circumstances. However, new formulas are likely to be required to deal with the increased pressures upon resources that are being experienced in so many places.

Rights in trees do not always accompany rights in land upon which they stand. In the western provinces of the Sudan this is the case for gum arabic trees. If this were more widely recognized as an institutional possibility and trees seen positively as an economic resource, then poor people could be given title to trees, or use rights in trees on public or common land regardless of their own landholding status. In this way incentives can be provided for increased cultivation of trees while protecting the interests of the poor. This is an emerging practice with *bor* fruit trees on village *panchayat* land in the state of Gujarat in India.

Elsewhere a landholder's rights in trees on his or her land are constrained by the state or other interested parties. A common environmental protection measure has been for a prohibition on the felling of trees. However, it is now apparent that this provides a disincentive to landholders to plant trees, since the economic benefit that they could obtain is limited (Foley and Barnard 1984). More constructive are rules which require the replacement of trees that are felled.

If there really is a commercial value to trees for fuel or building purposes, then the privatization of forest areas on a peasant-farm scale can be contemplated. The land is likely to be kept under trees if the value of the tree crop, despite much longer maturation, is considerably greater overall than the alternative use of the land for cereals or other crops, though care would need to be taken to see that the farmers do not suffer serious cash-flow problems and that reasonable allowance was made, through intercropping and other measures for subsistence needs.

Communal wood lots, now being attempted in many countries, at least on an experimental basis, depend entirely upon the creation of an appropriate institutional form. Here the problem is not only to allocate land but to set up the expectations of benefit so that the people will support the allocation of land for this purpose and allow the trees to mature. Equally, in the real world, there needs to be the expectation of costs or punishments if attempts are made to make non-agreed private gains (Shepherd 1985). In countries such as South Korea, which have made extensive progress in community forestry, the state has had a heavy financial and organizational involvement. This may be expected to be necessary in other countries too (Foley and Barnard 1984).

Most large forests remain in the hands of the state. People who have customarily lived in these forests may be entitled to pursue some limited agriculture and to the use of certain specific forest products, but the lion's share of the value of the forest production is supposed to go to the state. The state protects its interests by the employment of forest guards. In practice, some of the value of harvested timber will often 'leak' away from government, but more often to contractors than to forest dwellers or forestry workers. This situation should encourage an investigation of improved incentives to forest workers or users by entitling them to a better share of the benefit of growing trees. The *taungya* system – started in Burma – under which peasants clear forest land, plant trees on behalf of the Forest Department, and are entitled to intercrop on a temporary basis on their own account, is but one of several possible arrangements for sharing costs and benefits in an organized manner (CERES 1985).

Water

Water rights in arid places are more important than land rights. Here we will only discuss water for livestock. Water rights are equally varied and sophisticated as rights in land, but fairly common legal solutions seem to be that where nature provides the source, access is unconstrained (for those people entitled to pasture animals in the territory) but where a well or a *hafir* is dug, the water rights belong to those who undertook the construction.

In the past, individuals or groups dug most wells and present generations have inherited the rights and obligations that go with them. Nowadays governments, sometimes backed by donors, are taking a larger role in water provision. In principle, this should put governments in a position directly to control who can use the water and under what terms and conditions. Again in principle, water can be controlled by governments in another way also, that is by requiring permission for a water source to be developed. In practice, it has proved extremely difficult either to control water use at government-owned water points or to control the spread of privately owned sources.

The main problem has been the very attractiveness of water-supply investments. It sounds right to invest in water in arid lands. But for governments it is also a means of gratifying personal or group pressures within the political system, and for donor agents the supply of drills and bulldozers is an easy way of meeting the spending imperative to which all development agencies are subject. Sometimes the objectives have been to control over-exploitation of grazing by preventing the over concentration of watering points, but as a Sudan case study illustrates (Shepherd, Norris, and Watson 1987), the planning and administrative machinery to achieve this end is often simply not available, or is up against seemingly irresistible pressures from the politically powerful. The other objective may be to provide water to the poorer stockholders. This may be more readily achieved but to small advantage if the surrounding grazing is destroyed.

The discussion of water illustrates the point that development problems are more often problems of control than of money. And problems of control are particularly difficult to solve when the rich and powerful are in a position to object. However, overgrazing is in nobody's interest, and it should be possible to find a better set of institutional structures for water development than exists to date.

Grass

This leads into a discussion of grazing which has been touched upon several times already. The inequity and sometimes technical inefficiency of privatization of grassland has been discussed. Group ranches, attempted in Kenya, have proved more attractive in the protection of graziers from encroachment by cultivators than in achieving land-use control objectives (Grandin 1986). Where no change in the legal status of common grazing land has been made, two kinds of constraint can nevertheless be attempted (apart from indirect control through regulation of water). Access can be limited to particular groups of people, perhaps at different times of the year. Where pastoralists are better armed than officials, such regulation must necessarily be by mediation and agreement between parties. Attempts are now being made to restore this tradition in the Sudan after a period during which disputes have tended to move into the political arena and remain unresolved.

A second, more difficult measure is the limitation of entitlement of access for graziers to particular numbers of cattle. In colonial days the cattle cull was introduced occasionally to reduce numbers proportionately in all herds in times of environmental stress. More rigorous (and perhaps only suitable in essentially arable areas) is the permanent limitation of stock-holdings of each resident in an area, as is the practice in parts of the British Isles. Such measures are not ones that politicians like to promise to their people, but the people might be prepared to impose restriction upon themselves if decisions rested with some locally collective body and benefits came to them and not to others.

Credit

There are contradictory views in the development business about the role of credit. For some it should be made available only with caution, while for others it is still regarded as a panacea of all ills (Famine 1985). Nowhere in Africa has credit become so prominent a development instrument as it is in India, where the major anti-poverty programme is based upon it and where the banks are obliged to lend a proportion of their overall portfolio to agriculture and to the poor. Nevertheless, it is an important resource and the rules of access to credit, which normally favour the well-off, can be made to accommodate the poor and contribute to their resilience. The need for collateral in the form of land can be replaced by mutual guarantees by groups of borrowers or by lending against a future harvest. Lending institutions can build in an element of insurance if they are prepared to write off loans in times of natural disaster. But only if they actually manage to reach vulnerable sections of the population and devise financial means of sharing risks with them, will lending institutions really begin to play a role in providing a substitute for the informal mutual-aid entitlements which farmers arrange amongst themselves.

Labour

Labour is increasingly a commodity in Africa as elsewhere. As such, it is sold in the market. There it may, in principle, be subject to minimum-wage legislation and other controls, though these are notoriously difficult to enforce. Commoditization is not going to be reversed, but there are still plenty of examples of labour-exchange arrangements and labour commitment to communal endeavours to illustrate that a different approach to the value of labour co-exists in Africa today with the labour market. In Botswana, we have seen, widows or the elderly still have the right to call upon relatives to plough their fields for them, and this is but part of a network of entitlements and obligations involving labour. But in this area as well, the pressures that develop within the rural economy can lead to a breakdown of these entitlements, leaving some categories of household particularly vulnerable. People without work is the more commonly recognized category of vulnerability; people without labour is another.

Entitlement measures that can go some way towards tackling the joblessness problem include settlement schemes and other means of granting title to land upon which personal labour can be used, job creation in other parts of the economy, as well as 'make work' rural works. We are now suggesting that 'make work' should, where possible, be accompanied by the granting of title of the assets created to the workers involved.

Vulnerability through labour scarcity can be tackled by legally reinforcing the entitlements that people have to the labour of others, by granting subsistence loans (which are also production loans) to families whose field labour would have to migrate elsewhere without such assistance, and also by enabling public funds to be invested in the improvement of the private productive assets of needy groups in the population.

The question is, How do these structures translate into institutionalized entitlements? Two things are required, one is a decision-making framework and the other a policy process.

A decision-making framework

Decisions certainly need to be made at national level. We assume here that basic ideological commitments are often immoveable – though there have been some spectacular reversals in Tanzania and elsewhere. Basic commitments still have plenty of scope for national bodies to make policy on such matters as conservation of land resources, the development and control of water, or enabling powers for local authorities. These may have a profound impact upon entitlement systems by providing a backing in national law for grassroots endeavours to establish controls. What may be more innovative is our belief that a local level decision-making and enforcement framework is also essential. Some such capacity is present in many countries. Botswana has its Land Boards as adjuncts of its local government bodies, and village assemblies are still important decision-making bodies. Sudan has provision for bylaws on land use and other regulative matters in its area council powers, though these are seldom used. Equally important are problem-focused mass meetings.

Local decision-making and enforcing capability is essential because:

1. Very often it is only at this level that essential knowledge of existing entitlements is available, often in people's minds.
2. 'Fine tuning' needs to be a local matter.
3. It is only in a local arena that 'small' people, be they women or less-influential men, can hope to be able to exercise their entitlements. In this case, though, there are two provisos. The first is that decisions are taken in public so that they are exposed to public scrutiny and criticism. The second is that central legal authorities are prepared to treat appeals from politically weaker citizens sympathetically.

A policy process

We can now consider how, at central and local levels, sensible and sensitive decisions about improved entitlements and obligations can be arrived at. Basic procedures can be:

1. Decide upon public policy objectives. This can be initiated at both central and local levels and the results drawn together for debate.
2. At each level, identify the various parties that have an interest in achieving or frustrating these objectives and establish what these interests are.
3. Work through these objectives and interests to come up with a formula that will both advance the objectives and take account of the interests. This will be an interactive process in which initiatives can come from either central agencies or local bodies.

This is a prescription for a public policy process that responds to the strictures of Grindle (1980) and others who recognize that administration and public policy operate in a political environment. All aspects of policy, from initial drafting through to design and implementation, and concerning investment decisions as well as institutional development, need to be put through an interactive process of this kind.

The strength of this approach should be that it recognizes that needs can be perceived from a local as well as a central viewpoint and public as well as private interests may be involved. There may be conflicts as in the case when a public authority wishes to take over a piece of privately cultivated land to make way for a road or a school. But, equally, private interests may be furthered by the successful protection of public interests. Where this is recognized, public policy can be moved on from an 'either/or' choice and away from oversimplified ideology-based solutions. It is in the development of *thought-out formulas* , recognizing complex and multi-faceted interests, that the hope for sound legal institutional development lies.

10

The Roles of Government, Aid Agencies, and Local NGOs in Famine Prevention

In this chapter we pull together a number of the findings of earlier chapters about 'counter-famine' strategies and assess the capacity of governments and other agencies such as non-governmental organizations (NGOs) to carry them out or to select suitable strategies. The success stories from Botswana and Gujarat (see Ch. 5) have a common emphasis upon the importance for famine avoidance of administrative capacity, sound finance, and democratic institutions. The chapters on emergency measures, livestock, and technology (see Chs 6–8) tend to assume a similar framework as the norm within which particular prescriptions can be developed. Yet, as our studies of trends in the Sudan and in Ethiopia illustrate (see Chs 3 and 4, respectively), in most countries one is dealing with a complex and often rapidly changing scene. Many countries do not share the governmental characteristics of Botswana or Gujarat, and there is clearly no point in simply prescribing them as a *sine qua non* of famine-proof development.

So in this chapter we approach the real world as it is. In asking the question 'How is it to be done?', we look for strengths within administrations that in conventional terms might be found lacking, for means of funding the otherwise bankrupt, and ways of achieving some sensitivity and responsiveness to the vulnerable social groups where formal democratic institutions are lacking. In doing so we have to consider the roles or potential roles of a range of agencies that operate within the rural scene. However, the current debate centres upon the role of formal government, so there we start.

Government should be assessed not in terms of abstract notions but in terms of ability to get things done. The things that need to be done to counter famine are many and various, as earlier chapters have demonstrated. It is quite likely that some governments may be better at some than others. Yet sweeping judgements and fashionable notions often prevail over pragmatic assessment of capabilities.

If we conclude that governments, assisted by aid agencies and watched by the international community on its television screen, have a poor record in the prevention of famine in Africa, we must ask what sort of changes are needed to improve that record and save lives. We have drawn attention to a range of policy measures in this book. In general, government and aid agencies should restructure themselves in order to focus more clearly on the problems of famine-prone areas. This sounds obvious but needs to be repeated, since policy formulation is often weak and famine-prone areas are usually economically marginal with problems that are daunting. The question that we address in this chapter is, 'Where will government, voluntary agencies and the aid industry find the capacity to tackle these seemingly intractable areas and problems?'

Needs and capabilities

The capability of the public sector, resulting as it does from a mixture of political processes and administrative strengths and weaknesses, is a very real constraint upon policy choice. It is not simply that some policies are unlikely to be adopted by particular regimes but also that which policy is actually best itself depends upon capabilities.

In the preceding chapters a number of technical and administrative possibilities have been discussed for dealing with the related problems of drought and famine. These are wide-ranging and require a number of different skills and capabilities. Drought preparedness entails monitoring such things as nutritional status, or the state of the crops or of the grain stores. Buffer stocks require procedures for their purchase, storage, release and disbursement. Equally necessary is a capacity to organize public works or to stimulate assisted self-help activities.

The demands upon administration can be summarized. There are things to do with information. There is the handling of goods; the collection, storage, and distribution of grain and other commodities. There is the management of people on site, digging dams or whatever. There must be assessment of local needs and priorities. Quite a different kind of activity is research and development, which even in industrial firms requires a separate branch of organization with different working procedures. There are the problems of commons management, partially considered in Chapter 9 but the public authority implications of which need to be considered here. Then there is the issue which we have labelled freedom from strife into which many African governments, despite difficulties which are apparent, have invested much thought and effort. With this we should start since much else depends upon it.

Notice that this list is not about projects. Projects are creatures of foreign aid and as such probably won't go away, in the near future at least. How to handle projects is a question which requires separate treatment, but the point that emerges here is that the real issues of drought management, perhaps development management as a whole, are those which concern the functioning of the politico-administrative system into which projects are injected.

Conflict management

Development in Africa, as elsewhere, has proved to be highly divisive. Partly due to cash-crop/food-crop price differentials, partly due to the habit of investing aid monies unevenly, and partly just as a result of commercialization, some areas, some tribes, some classes have benefited more than others.

Political responses have of course varied, but all regimes, even the military, have had to look for some basis of support amongst the population, and usually it has not been enough to rely upon the prosperous. Several mechanisms for handling local political pressures are now familiar. Area-based representatives of government, the chief officers of the district or region, often have an expressly political as well as administrative role. It is their job to harmonize political and administrative processes at their level, and to communicate with the national leadership. When they are given scope to get on with it – an important proviso – they can often be rather effective.

Another device is the containment of political forces within a single party. Political scientists have tended to look askance at this move because it does not sound democratic, but where an endeavour is made to pull in various factions and interests and come to a resolution of differences within a party framework, this can be a harmonious solution.

Local government is clearly seen by central political authorities as a device for allowing scope for the expression of local interests while containing them within a legitimate but manageable framework.

Containment of the scope of the political agenda – an aspect of what Kasfir called departicipation – is another device (Kasfir 1976). At best this is the result of a genuine consensus on objectives in development, at worst it is a result of a central monopoly on decision-making and control of the media leading to peace only in the sense of a sullen acceptance among the masses of the realities of power. But, usually, while some areas of policy are removed from public debate (and action), others are allowed to remain. Looking at the process constructively, one can see that there is scope for leaders to choose the agenda of policy debate and include items around which a consensus can be built.

These four devices are for dealing with divisive forces within an area. They contend with forces which would rather choose supporters and support them against the rest. Each device can in fact become suppressive, and most governments are accused of using them in this way from time to time. Dictators, military rulers, or ideological regimes of the right or left which rely upon foreign backers, use political administrators, the party, control of the agenda, and a tightly contained local government as a means of patronizing friends and silencing opponents most of the time.

In a similar way, one can see evidence of extensive attempts to handle inter-regional disparities; formulas for achieving regional parity in the distribution of the national budget, or more often parts of it; the appointment of ministers as champions of regions (as well as departments); and other measures. But again the weakness of such populist measures is that certain regions or interests, perhaps cattle people at the margin of an agricultural population, may simply not have enough political clout to figure in the power equation, and take to guns and raids instead.

So instruments for building consensus are usually in place, but the fact that large geographical areas are in conflict and many people are alienated attests to their weakness in practice. The most serious consequences of severe departicipation have been war and famine and sometimes both. It is no accident that the major wars and the major famines of recent times have occurred in countries with suppressive regimes. Democratic institutions are possibly the best safeguard against famine and secessionist conflict. In a democracy, whether one or multi-party, there are mechanisms of accountability which are not restricted to small groups of the population, so political structures will be critical and needs will be expressed. Ruling groups are both kept in touch and prevented from the most conspicuous forms of profiteering from the effects of famine and disturbance. Aid donors who are really concerned with famine avoidance, can be expected to exercise leverage to try to build up accountability procedures. They should look at the effects of their own procedures also.

To be effective the accountability instruments require a measure of decentralized power. However, there is a strong tendency for real decisions about policies and investments to be taken in central offices, in distant aid agencies, and by consultants, bureaucrats, or political ideologues in capital cities. When local bodies are told what to do, and it is as often as not wrong or inadequate in local eyes, it is unlikely to encourage political participation.

The alienation of local populations through ineffective decision-making institutions, has several consequences:

1. Responsibility for living harmoniously with neighbours whose interests may differ is effectively removed.
2. Leadership can take the form of a scramble for the spoils of 'development' uninhibited by the need to share them equitably among competing interests.
3. Local opposition to extractive projects or inappropriate programmes is defused.

The implication of this discussion for the role of government in achieving drought and famine-resistant development is apparent. Avoidance of out-and-out conflict is a precondition of the kind of administration that can monitor nutritional status or run supplementary feeding programmes. So a heavy emphasis needs to be placed on procedures for achieving political harmony at different levels. This sounds platitudinous and many people will point to the structures in place and say that all that can be achieved has already been achieved. However, whatever the structures, there is very little evidence of the kind of decision-making activity that would indicate effective local power and that authority is being allowed or exercised. In very few places can one find a locally worked out development plan – other than a shopping list of welfare proposals. In few places has drought or famine occasioned a council or conference, with minority group participation, to negotiate land use or drought-prevention measures. In few cases have aid agencies put aside pet projects or commitments to put money up front for locally determined priorities. And in most cases ministries have reasserted existing commitments as determined by headquarters without allowing for local discretion in their selection or implementation. In these circumstances there is not much scope for generating an area-based consensus.

To put this positively, what we are saying is that besides means of making leaders accountable, people need to be able to influence decisions about their environment. Direct means of participation through interest-group organization, mass assembly (considered further below), organization of the poor, around specific environmental issues or investment opportunities, should be tried. This may be possible even when formal democracy is not.

Let us discuss this also in terms of its effects upon social structure and inequality. We have identified land grab and the subsidized access to capital that is often associated with it, as part contributor to the vulnerability of rural populations to the effects of drought, and often as substantial contributor to environmental degradation. The accusation is often made that decentralization can lead to control being handed over to local élites who are thus enabled to legitimate their privileges. This may indeed be the case where an emerging capitalist class faces disorganized or fragmented peasant or pastoralist interests. However, many of the schemes, like those on the clay plains of Sudan, through which capitalist agriculture is being advanced, are centrally sponsored and not subject to any local controls. Had they been subject to any effective popular decision-making at a local level, they either would not have taken place or would have been substantially modified to reflect the interests of – in the Sudanese case – the pastoral peoples whose seasonal pastures the clay plains once were. So decentralization can be a means of exercising restraint upon exploitative and unbalanced development if relevant interest groups or classes have an effective voice.

To have an effective voice, procedures would have to be open to participation by the masses. This entails submitting all important issues to debate and decision in mass assemblies. Elite councils, appointed, hereditary, or elected, normally will not do because any procedure which allows leaders to take decisions behind closed doors is an invitation to exploitation and collusion. In mass assemblies leaders have to make their points in the presence of their followers as well as their rivals and cannot be seen to be promoting narrow personal or sectional interests. The councils once held for the settlement of disputes amongst nomads in the Sudan are a case in point. At these councils opposed groups would assemble under the chairmanship of the administration and take decisions about land use and other matters. It could be a long, drawn out affair, but it led to some control of resource use as well as of social relations.

If consensus-generating procedures were given scope, we are

confident that many drought and disaster prevention measures that we have discussed in the technical review chapters would in fact emerge as things that people want and agree to.

What should be the role of other agencies in conflict management? The principal responsibility clearly belongs to government for keeping the peace, but given that conflict may well result from the under-representation of groups or interests, there may well be a role for voluntary agencies, religious organizations, even economic organizations like co-operatives, to ensure that minority voices are heard in the public decision-making arena. On the other hand, outside agencies such as foreign NGOs, for all their talk of 'conscientization', 'participative organization', and the like, usally steer clear of involvement with minority representation for the obvious reason that they are there on sufferance of the powers that be. This being the case, the main responsibility of outside agencies may be to ensure that their investments and activities are not in fact themselves divisive and conflict-generating. This requires a consciousness of social structure and interests which project planning procedures do not in general supply.

Commons

The need for consensus-generating procedures is particularly apparent in the case of commons improvement. As we have seen in the previous chapter, commons improvement involves constraints. The ability of most governments to constrain pastoralists or even agriculturists with livestock on the commons is minimal and is unlikely to increase. Similarly, a government's ability to prevent people cutting down trees is almost nil. So improvement measures will depend largely upon self-enforcement, which in turn depends upon agreement between interest groups. The role of government is first to establish the legislative framework within which agreements can be arrived at and enforced and second to provide an agent of central authority who can encourage the process and sometimes hold the ring until conflicts are resolved. Thereafter a local authority of one kind or another needs to take on the responsibility.

This is very much a public-sector matter and should be recognized as such by other agencies whose main responsibility should be to avoid setting awkward institutional precedents by demanding special concessions or establishing 'pilot projects' which depend upon setting up privileged access to what would otherwise be common property.

Activity management

By activity management we mean the organization of feeding programmes, building activities, digging trenches for water pipes, or whatever else is necessary at village level. This begins to call upon more conventional administrative and technical skills. Public works programmes and the like require a list of approved projects, recruitment procedures, project supervision, and financial controls. However, if engineering and administrative skills are in short supply as is sometimes the case – though the more common condition may be lack of personnel and shortage of funds – there are ways of setting up procedures that call upon locally available skills and resources. For instance, it is common for village development committees or similar bodies to be involved in project selection, a procedure which often ensures that projects are put forward which are also within the organizational capacity of village leaders.

The problem of selecting needy persons for places on public works has sometimes been put as a challenge to village assemblies, trusting that the pride and paternalism of the better-off will lead to preference being given to the poor (also assuming their willingness to pass responsibility for maintenance of the poor to the machinery of the state). This was done in Botswana during the drought of 1969–70 quite effectively. In village assemblies teams of workers for a Food-for-Work programme were selected without difficulty, though the practice emerged of selecting a second team to alternate with the first on the grounds that a larger number of families than could be accommodated on one team were in need (personal observation, Donald Curtis). This practice, then a widely used subterfuge, is now an established principle, as our Botswana case study shows. In some Indian states, under the so-called 'Antyodiya' programme, the practice of asking villagers to identify the poorest of the poor for special assistance is also applied, with some success. In these ways grassroots capabilities are used to the full and the burden on administration in the district or region somewhat reduced.

If engineers or other specialists are available, one way forward may be to get them to act as consultants to local decision-making bodies rather than to expect them to design projects in their offices, so leaving the initiative and responsibility at the village level.

By drawing upon local initiative and organization, the burden of drought administration is shared but not removed. In fact, the organization of participation needs special skills of its own, particularly political awareness. This is something which seldom features on training courses but fortunately is widely available in many parts of Africa and can even survive the pomposities of administrative training.

Participation also requires some kind of popular movement if it is to have meaning and substance. And a popular movement implies political leadership. In many parts of Africa there are or have been strong movements of this kind around 'self-help' activities or other aspects of community development. Most self-help activities have been about schools and clinics where the danager is that popular enthusiasm overcommits the public sector to provide teachers, nurses, and materials. A further danger is that political leaders use project organization simply as a means of enhancing their own status while providing patronage to their clients. Nevertheless, many of the lessons of co-ordination of village-level activities and public-authority budgets have been learnt in countries like Malawi and Kenya quite well enough to make self-help a significant element within rural development.

Self-help can also be applied to drought-proofing and soil-conserving measures. In fact, the greatest amounts of earthworks anywhere in the world have been done by farmers themselves working as individuals or in small co-operating groups. Self-help schemes could be based around the provision of central resources to farmers who group themselves into mutual-aid teams on projects selected by themselves.

Public works are of course rather different, not least because they establish a different set of expectations amongst the rural population as to what benefits they will get from taking part in community-based activities. Nevertheless, where beneficiaries do have a voice in deciding what is to be done, or if it did become the practice to link participation with long-term benefit (Ch. 9), public works can easily become an instrument of a social movement. In these circumstances the element of patronage and protectionism that is an unavoidable part of the political process anywhere, can be turned to good effect. If land improvement and water conservation are the main projects undertaken, this does not give much scope for the creation of follies to stand in memory of some politician. The distribution of food or wages in return for the creation of long-term productive capacity is not the sort of expenditure commitment that easily gets out of control.

Routine services

Veterinary or health services, education, agricultural extension, or the running of public works departments can be regarded as routine services of government. They are important to the present discussion both in their direct contribution to resilience and as a base for emergency programmes.

In earlier chapters we considered the relevance of various programmes that may be run by routine service departments, to famine prevention. Here we must assess the varying capabilities of such departments under prevailing circumstances.

All the activities discussed in earlier sections of this chapter can thrive in a lively, participative, political atmosphere. More difficult are the routine services, because here long-term pressures build up (1) for expansion of employment, at the expense of other items on the recurrent budget; (2) for promises of universal coverage which cannot be met; and (3) for favours that are difficult to deny behind the closed doors of bureaucracy.

So we find vets without medicines, service plans shelved in district offices, and grievances explained in terms of favouritism to others even when scarcity itself is an adequate explanation. These tendencies, which can be found in many countries in differing degrees, are both consequences and causes of political processes. They follow from the need for politicians to make offers or promises in order to recruit support. They lead to the situation in which promises can seldom be fulfilled and become bad currency. Politicians then retreat behind administrators, whose main quality must be that they are able to resist grassroots pressures. Administration in turn becomes 'unresponsive', though it may still be open to persuasion to grant particular favours to the rich and powerful.

These tendencies have by no means been universal. There are examples from many countries – Kenya, Tanzania, Botswana, and elsewhere – of well-maintained services that have been supported rather than undermined by politics. However, service budgets have also been subject to pressure from above. As world recession, falling export commodity prices, debt-servicing requirements, and the rest, have hit at national level, the amount of money that comes down through the system to pay salaries or provide essential goods has dwindled. So services are caught between shrinking central allocations and grassroots demands, expectations, and commitments.

The consequence is that in many places now – particularly in marginal, drought-torn lands – rural services have withered away, leaving underfunded and sometimes overstaffed activities in large villages, towns, and district capitals. The costs of this state of affairs fall upon the consumers who are made the more vulnerable to the risks of famine. For example, in arid lands families may have to spend one-third to one-half of their income on drinking water because of broken-down pumps or failed supplies. Distress payments for basic services like water reduce the scope for any form of saving or productive investment to provide for lean years.

The condition of remaining services depends upon what the popular demand is for the service and the degree to which its provision depends upon a reliable supply of scarce goods such as foreign manufactured drugs, spare parts for vehicles, and petrol.

Education – probably the least relevant to famine prevention – is relatively robust. Teachers tend to live alongside their schools and parents want their services badly enough to be prepared, in some cases, to feed them when their salaries do not come through or to employ them directly if the authorities are not able to do so. A division of education services into formal and informal or 'recognized' and 'unrecognized' is common. The informal or unrecognized could be seen as a form of private solution to service provision that arises out of local demands and long pre-dates the current fashion for privatization within the development industry. However, much of it is organized on a community basis and is not run for profit or private gain. It is attractive in that people actually get what they pay for, though standards may be modest and staff may suffer insecurity and low pay. So, one way or another, demand keeps some education going. In the face of scarcity of central funds people find ways of raising local resources. What may be needed is for authorities to find more effective ways of incorporating these into the provision of routine service (Curtis 1988).

Human and animal health services – highly relevant to both famine anticipation and prevention – are much more vulnerable due to their dependence on scarce supplies. The great planning era, effectively the 1970s in most places, saw the production of elegant documents setting out maximum walking distance criteria for the location of primary treatment centres, establishing the principle of prepacked supply kits and routine servicing. A lot was achieved both by highly committed governments like that of Tanzania and also through programmes sponsored by international or voluntary agencies and backed by their funds. However, any interruptions in the flow of supplies destroys the credibility of the system. Illegal privatization, that is the diversion of such supplies as turn up onto the market (and often into unqualified hands), can follow.

Under these circumstances services may break down. This is often in fact the point at which foreign funding bodies intervene. Take-over by agencies run or funded by foreign donors is becoming an increasingly common solution to the problem of inadequate services. It leads to a kind of 'neo-colonialism' in which foreigners impose policy as well as influence day-to-day management. This situation is justified in project documentation by labelling it 'transitional', designed to restore levels and standards of service within a short term. It is often associated with an emphasis upon training and use of counterparts for foreign contract personnel. In these ways the transitional nature of the intervention is emphasized. However, of itself this formula does not necessarily generate the indigenous resources that will, at some future date, be able to replace the foreign resources. Foreign intervention may create a non-viable precedent. And in many cases the problem may not be a lack of knowledge and skills – though on-the-job training is always a necessity – but a lack of willingness or capacity to raise the necessary local resources and to maintain an adequate budget for non-salary recurrent expenditure requirements.

We return to the question of what foreign funding bodies may do to avoid creating dependency. But it is worth considering the possibilities of reform within service provision by government agencies because it is all too easy for foreign funds to be a substitute for such reform.

Reform opportunities in routine services provision

The great endeavour of the independence period in many countries has been the provision of universal basic services that are free at the point of delivery. This kind of commitment has been equally asserted by regimes of the 'free market with planning' variety as in socialist regimes. It has even been a commitment in Malawi where planning itself is rejected. This is the kind of commitment that is very difficult to get out of, should one want to, even though, at anything short of 100 per cent efficiency, the system leads to great disparities between those who get the service free and those who do not. It is very difficult to prevent those who are in any case privileged from getting prior access to available services. In most cases, also, a market develops at the boundary of the free service to make up for its shortcomings or exploit its weaknesses.

A reform might manage to get a wider provision through making education nearly free instead of free, or drugs at fixed prices and available instead of free and non-available. However, resistance to this kind of notion is strong. It would, for example, require a constitutional change in Sudan. It would necessitate very careful political presentation in most places. It is also something that would have to be negotiated through a number of informal as well as formal interest groups because many people have developed interests in exploiting the shortcomings of the free service. From the perspective of villagers, it would make regular and control a situation that already exists. Most services at the village level are not in fact free.

Pressures for reform are building up, not least from the leverage which external funding agencies are able to exert in countries with severe debt problems. In many cases such pressures have not as yet led to positive reform. It appears that the financiers are prepared simply to put the pressure on for stringency measures without specifying how public-sector economies are to be achieved – other than general praise of the virtues of private enterprise and capital. Instead of straightforward privatization, a number of other reforms are possible, depending upon what are the most pressing problems.

If improved revenue is a prime objective, as it often will be, and domestic rather than foreign resources are to be used, higher taxation revenue or more income from fees and charges are the two possible routes. Substantial improvements in revenue should be possible in many places. While dramatic impoverishment results from drought, it should not be assumed that rural people, particularly livestock owners, are too poor to be able to pay in normal times. Often people are in fact already paying substantial sums for poor quality private services while avoiding payment of taxes. What may be required is a much closer link between payments of taxes and tangible benefits from such payments. Too often taxes raised in rural areas have gone to pay urban salaries with little or no benefit being transferred back to the rural areas. Then tax collection is reduced to predatory raids by officials upon those in the rural population who cannot run away.

If people have to pay for a service at the point of delivery, at least they are sure that they are getting something for their money. So service charges are one way forward. They can be flat rate at average cost, or differentially charged with payments for higher levels of provision subsidizing more generally available lower-level provision (as in some domestic water supply systems where payments for house connections can subsidize standpipes), or organized in innumerable other ways to make them reasonably equitable and politically presentable. There should of course be a reduction of general taxation to offset such charges.

We are aware of Leonard's argument against user charges, that farmers are in any case paying a large hidden tax in the form of lower prices for export crops than their foreign exchange value warrants (Leonard 1985). Where this applies there is certainly an argument in favour of free services. But it is not always true, and even where it is, it will seldom be the case that the revenue so raised accrues to a 'rural services account'. Our point is that to be able to withhold payment provides rural people with a means of exercising leverage against the school teacher, the nurse, or the vet if a service is not provided. The payment of a local tax provides no such sanctions.

The next problem may be to get an adequate part of improved revenues into funds that will not be subject to pressures for ever-expanding employment. In the education sphere the people's own solution to the problem seems to be to place increasing reliance, for books, for equipment, for school meals, and other non-salary recurrent costs, upon a 'voluntary' school fund. This is a level of decentralization that is seldom officially contemplated but has its attractions. A school fund, if it is not to be 'eaten' by hungry headteachers has to be subject to popular control at parents' meetings. Where this is achieved, the use of the fund comes to reflect locally determined needs and priorities also. Another, perhaps complementary, approach to the problem is to get the professional interests of the providers of public services involved. Medical or veterinary staff have an interest in ensuring that they have medicines in stock and this could be the basis of an arrangement for their management of a fund outside normal government or local government revenues, subject to some form of counterchecks.

User charges cannot be applied for services that benefit people as members of a community and not as individuals. Tsetse fly control is one example. All members of a community benefit in terms of their own health and that of their cattle, but no one has an individual incentive to pay for it. In colonial times compulsory labour, to clear the dense bush that harbours the fly, was enforced. Today only the compulsion of a taxation system can provide for modern treatments. Common provision of food reserves is also in this category, and there will be a range of other provisions that simply cannot be met unless a public sector capacity can be maintained.

This leads us to discussion of the problem of motivation. The generalization would probably hold for many parts of Africa that there are good, well-qualified people at many levels of the public service who are extremely frustrated, not only by poor pay and conditions of service but by inability to do the job for which they are trained for want of necessary resources. Sometimes foreign funding agencies, finding government badly managed, overmanned, and undermotivated, are setting up alternative structures and attempting to recruit qualified staff by paying higher salaries. This has been done on a truly massive scale by the World Bank-sponsored Agricultural Development Projects. This can deprive the public sector of the best of its personnel and compounds the long-term structural problems that emerge when the foreign funds dry up. Far better to tackle the problem of public sector motivation head on.

Aid agencies have tackled the problem of motivation by money incentives in the form of salary top-ups, *per diems*, and bonuses to officials involved in aided projects. In Sudan this has reached the stage where officials' salaries are being doubled in many cases. This has many negative consequences: government-funded programmes do not attract the most able or ambitious civil servants; competition builds up between aid agencies over the rewards that they build into their projects. However, the prevalence of the practice makes clear the critical state of civil service pay and may, in the medium term, lead to a more satisfactory solution to the problem. In one case, that of a UNICEF funded drinking water supply programme in Kordofan, Sudan, the government has itself adopted a system of productivity bonuses which was first introduced by the aid agency.

This case illustrates the point that motivation can be much improved by productivity bonuses. After introduction of this scheme, annual installation of water pumps improved by 200 per cent. Where a project can be managed on a cost recovery basis it could even be consumers who pay all or part of the bonus.

It is not possible to come up with universal prescriptions as to how to improve motivation beyond the most general. Heaver points the way with the suggestion that administrative systems need to be designed or redesigned with the interests of all parties in mind (Heaver 1982). This follows the observation of Chambers that, as often as not, when you explore the situation of junior officials, you find that they face disincentives to getting their jobs done (Chambers 1974). For instance, field staff may be out of pocket if they actually visit the field. (It works at the other end of the system also. Development 'experts' may enjoy higher *per diem* allowances if they sit around in capital cities than if they go anywhere near the scene of development activities.) Advance in this area can only be by vigorous and probably painful re-examination of the innumerable rules and regulations that go to make up terms of service.

The objective – if it is taken seriously – should be to become as clever as the private sector at motivating people to do things. One thinks particularly of the way in which merchants' lorries are kept going into old age by their drivers while relatively new government vehicles litter the forecourts of government offices propped up on bricks. Merchants' drivers often handle large sums of money for their employees as well as undertaking repairs. Sometimes they work for a fixed contract for a journey and have an interest in making the journey as inexpensively as possible. Sometimes they carry passengers on their own account to supplement wages. Often the reward system that is arrived at is of questionable legality (which may have results for the reward system of the police). Often the drivers do well enough out of the system to set up in transport on their own account (other kinds of workers may not have as much leverage). But it does lead to the conveyance of people and goods economically, and in that sense it should be emulated.

Is there the will to break through to reform of this kind? There is little evidence as yet to suppose that there is. But in many places rewards for service in government have become so low that civil servants find land and cultivate it, or get somebody to cultivate it on their behalf. Many officials at all levels have recourse to 'moonlighting' in some other kind of employment. These are conditions in which, from the employees' angle, there should be room for negotiation on the basis of increased rewards while from the authorities' angle there could be an insistence on formulas that increase effectiveness. The bigger problem may be that the very idea of drivers or teachers or executive officers being rewarded on the basis of productivity or being motivated to share in the maintenance of public property is strange. It does not fit in with any currently popular political philosophy; it does not go with the prevailing ethos, expressed in law in Zambia and no doubt elsewhere, of government as the employer of last resort for educated youth.

In sum, improved local revenues, a maintained budget for non-salary recurrent expenditures and built-in incentives would seem to be essential for any resilient service and should be the objective of reform in public administration. Virtually all of the policy changes suggested in previous chapters require hard work on the part of officials. Motivation may entail some changes in the terms and conditions of service as well as in principles of service provision. Emergency measures may need to superimpose external resources, but it would be a great mistake if foreign funds and agencies distorted the objective of improved government capability by being too ready to provide substitute funds and personnel in organizations which bypass government departments and duplicate their functions.

The NGO as an alternative provider of services

Indigenous non-governmental organizations in Africa fall into different types. There are the often very lively mutual-aid organizations such as rotating credit organizations or irrigation organizations. These are definitely providing services which are vital to the public they serve. The importance of this kind of organization has long been overlooked in the development literature. Such organizations are the product of need, whether this be for 'lumpy' finance, mutual assistance during life crises, managing seasonal peak activities, or in the organization of difficult tasks like bringing water to dry land. Such tasks are at the core of self-reliance and provide more for the basic resilience of a rural community than is ever likely to be socially engineered through the projects and plans of the development experts. While it is very difficult for outsiders to a community to have the empathy and social imagination to be able to stimulate such organizations when they do not already exist, experts at least need to know what exists in any social setting and make sure that plans and programmes do not replace such organizations with less viable substitutes.

Equally valuable in drought situations may be the kinds of association which feature prominently in parts of West Africa through which urban dwellers express their links with their places of origin in the countryside. These have been very prominent in raising capital funds for self-help projects in the rural areas and reflect a network of ties between city and countryside. Without evidence to substantiate the case it is to be expected that these associations can be mobilized for relief purposes when necessary, as is the case in India.

What has certainly happened during the recent famine, in Khartoum at least, has been the emergence of middle-class-led philanthropic organizations which took a direct role in organizing relief supplies. If this experience was paralleled elsewhere, then what we may be experiencing also is the beginnings of an organized interest group that is well placed to act as a spur to government activity: what is sometimes called a 'countervailing force'. The evidence to hand is based upon personal observation (AWS) and it is not clear upon what scale it took place. Many commentators are, however, looking to the emergence of countervailing forces as the most important factor likely to improve the performance of governments. Certainly, in Khartoum, voluntary action preceded official action and was influential in ensuring official recognition of the plight of western regions.

More prominent in many parts of Africa have been the foreign-based non-government agencies (NGOs), working sometimes in collaboration with national voluntary agencies or employing national staff in the field, but often extensively staffed by expatriates. Leaving aside the special circumstances of famine, when whatever staffing arrangements suit the needs of the moment are probably acceptable to all parties, one still finds many areas in which foreign NGOs act as virtually independent agencies, certainly under formal agreement with governments, but with very few day-to-day ties with the fomal structures of government.

The southern provinces of Sudan have been a case in point. There, until very recently, one agency, Norwegian Church Aid, employing substantial Norwegian government funds as well as its own (as is becoming the fashion amongst western donors) was, in the name of an integrated rural development project, providing virtually all available public services within its area of operation, except education and law and order. After some years, efforts were made to associate the provincial government with the work, but this government had less-qualified staff to act as counterparts to the foreigners and, more importantly, had little prospect of having the recurrent funds to be able fully to take over the infrastructure that was being developed. This was only one of several agencies in the region each of which tended to be better equipped with transport (including light aircraft) than the arms of government with which they were formally collaborating.

Agencies such as these are sometimes able to put more material resources and better-trained personnel into the field than national governments. They may find staff, who because of superior pay or as dedicated enthusiasts or romantics, are prepared to stay in places that nationals with equivalent skills are reluctant to commit themselves to: not just because of urban bias but also because they might get stuck there on lower and more unreliable pay than a volunteer and be missing out on career opportunities that are different from those of the foreigner. People in the areas so served may be provided with something that they otherwise would not get.

So NGOs can contribute genuinely additional resources of funds and manpower. However, in many places they simply represent an alternative channel of funds from foreign governments or international aid agencies which are happier to give money to a voluntary body than to a government. USAID and UNHCR have led the way in this. In extreme cases the funds are then handed on to units of the government which effectively then fall under the supervision of the NGO. In other cases the government seconds personnel into a project managed by an NGO. In these cases it has to be asked what the NGO is contributing beyond project supervision that is cheaper and perhaps more amenable from the international aid agency's point of view. These procedures may solve temporary problems, but they will not help to reform host government capability.

If NGOs are to make a long-term impact through their activities, more care than in the past needs to be given to the way that they relate to the setting in which they operate. Projects and programmes, even basic welfare projects, need to be founded on, and sustained by, a firm base in increased economic activity. There has been some commitment in this direction in the past in so far as projects have been submitted to some cost-benefit appraisal. But this has often been *ex ante* and notional and has not featured prominently as a determining factor in a project's continuing existence. What would be welcome would be project designs that require economic activity improvements to be built-in, perhaps clinics sustained by vegetable plots, schools by wood lots, and so on. Capital expenditure should only be incurred when there is a committed revenue source and an organization responsible for their maintenance.

If these principles were followed, a lot of wishful thinking about 'pilot projects' acting as a 'model' for future developments would have to be abandoned. Instead of thinking in terms of example – a paternalistic notion at best – there is a need to be aware of precedent, and to avoid setting false precedents, in terms and conditions of service, in standards of provision, and above all in what is sustainable.

The formula for implementation and transfer to local bodies needs to be worked out with the relevant authorities and with the local community at the beginning and reviewed en route to completion. There is always some formal agreement at national levels which governs voluntary agency activities, but this seldom leads to their being integrated at lower political levels.

The exception to the first and second conditions should be projects which are genuinely and expressly experimental, where a different kind of interaction with the host community needs to develop. To research and development we now turn.

Research and development and the intelligence functions of government

At several points in the chapters on emergency measures and technology developments, we have returned to the theme of a need for research. This is not just an academic's reflex reaction to a shortage of literature in any area, or a thin disguise for employment generation for fellow academics. Indeed, one of the problems with the suggestion may be that in many cases it is not the kind of research for which academics get their rewards that we are discussing. What we are simply pointing out is that there are considerable problems with getting good ideas, about cropping technologies, or fuel-saving devices, or whatever, from the mind of the programme officer, or indeed an innovative farmer, to the point where they fit the economy and ecology of an area well enough to be taken up on a large scale. The point to be made here is that this transformation, commonly organized under the heading of research and development (R & D), sets particular demands upon an administration which are seldom met adequately and which therefore come to act as a severe constraint upon what has been done in this area.

Recently, a lot of attention has been given to the question of what kind of intellectual process R & D should be and who should be involved (Biggs 1981). In the agricultural field 'farmer back to farmer' has become the catchphrase as we have shown. Organizationally, the implication of adopting this model is that R & D should take place in units that are located close to the farmers in question. Researchers should be in a position to have more interaction with their clients and collaborators than with their bosses. This requires the setting up of more or less independent agencies in widely dispersed locations with staff who are adequately motivated to get on with the job.

So the prescription goes, but in reality it has seldom been done. One problem, perhaps the main problem, is that the idea of R & D does not fit easily with 'development in a hurry'. People doing development programmes need to feel that they know what they are trying to advance. It is even more agonizing, in response to famine and drought, not to be able to come up with some ready-made technical fix. It seems to be almost cheating to have to say, along with Marshall Macluhan, that 'the medium [here R & D] is the message'.

R & D is also very difficult to control financially. There can be no direct assurance that what is put in in terms of funds will have a predictable outcome in terms of innovations developed. It can be a bottomless pit. Couple with this the fact that it is a very poor subject for political patronage – few voters or educated unemployed youth can be gratified by adopting an R & D strategy – and you have two more good explanations as to why it seldom has a high profile in development plans. But if the hard reality is that ready-made technologies seldom fit the almost infinite variety of rural environments, this has got to be faced up to. So we think that it is worth the while of anybody who is concerned with this problem to stick doggedly to the idea of setting up R & D cells wherever possible.

R & D cells do have some features which could facilitate their establishment. They are and should be small and left to manage within a fixed budget. They can be attached to a variety of different parent bodies, universities, or technical colleges as well as line ministries or local government bodies, depending upon who sees the purpose and who has staff interested in serving in them. For staff, they could be rewarding postings since any reasonably bright work done in them will lead to innovations that will present a challenge to the activities of other branches of government and a prominence to the individual responsible. Of course this is risky, but bright people advance by taking risks.

The voluntary sector, too, can usefully be involved in R & D. In Botswana, which has a large number of active voluntary agencies, much innovative work has been done over the years in farm technology, agricultural research, forestry, textiles, and other areas by cells set up and funded by these agencies. One advantage may be that their funding becomes relatively independent of politicial urgencies. A disadvantage, if NGOs have a foreign base, may be that priorities tend to be set by foreign enthusiasts whose links with and knowledge of the local economy are tenuous.

The idea of the R & D cell, like so much in organization theory, comes from big business. Recent work on large corporations in the United States is now changing the message slightly, stressing the need for innovative behaviour throughout, if a corporation is to be successful. And, of course, no sooner is this 'discovered' (which in the academic world means generalized in print) than someone, David Korten in this case, applies the remedy to development administration as well (Korten 1984). The message comes out of an empirical study of successful large American businesses which discovered a number of other things besides. The most important of these was that successful corporations had a strong corporate identity and sense of direction. Innovative behaviour could be seen as necessary in the face of a rapidly changing economic, social, and ecological environment, but it resulted from the response of highly motivated people responding to the challenge of the corporation that they served. Innovation often had to be managed out of scraped-together resources and in the face of inappropriate procedures (Peters and Waterman 1982).

Seen in this light, the necessity for innovative behaviour, which we have also identified in drought-prone rural environments, is not enough. There has to be a sense of purpose and direction as well. The very challenging task of making these environments more drought resistant might inspire, but then there are some difficult political equations to be worked out, innumerable donors with their separate demands and conditions to be handled, continuing problems with inadequate conditions of service. Korten might be happier if he had been able to cite studies which replicated the findings in development administration. As it is, the message sounds sufficiently convincing for it to be taken seriously as an objective.

R & D is an aspect of the wider business of intelligence. The monitoring or signalling of drought or other threats is another. If the word 'intelligence' conjures up images of spy systems, that is unfortunate because it is a far better word than 'information' in the context of handling data about rainfall, locusts, harvests, or financial outlays. Data need to be both collected and interpreted; which is where intelligence comes in. This is a learning activity and is only possible if those who collect the data also have some authority to act upon them. So in this case as well we are arguing for an element of discretion to be exercised at local levels. Too often in the recent famine, the necessary information was available but not read and interpreted for what it was.

So far we have emphasized the need for routine collection of data, such as measurements of crop yields or child nutrition, which can provide indices of abnormal conditions. Intelligence can also be more probing. Alert officials go out and find out for themselves. Impressions can certainly be misleading but can be backed up by requests for more systematic information through routine channels. Use can also be made of techniques of 'rapid rural appraisal' to back up first-hand impressions. The chances of this kind of intelligence being effective are greatest if everybody, from the minister down, knows that drought and other misadventures are an acknowledged danger and that alertness will be rewarded rather than punished. The point may seem obvious, but our case-study material shows that this does not seem to have been the case in Ethiopia or the Sudan in 1984–5.

References

Abdalla, A. and Holt, J. (1981) 'A study of the spontaneous settlement of nomads in west Sudan', London: International African Institute.

Abu Sinn, M. (1980) 'Nyala: a study in rapid urban growth', in V. Pows (ed.) *Urbanization and Urban Life in the Sudan*, Khartoum: University of Khartoum.

Ahmed, el H. (1985) 'Piling up problems: drought-famine aid', *Sudanow*, July, pp. 9–14 (Khartoum).

Awad, M. H. (1984) 'Economic Islamisation in the Sudan: a preview', Khartoum: University of Khartoum, Development Studies Research Seminar no. 50.

Badal, R. K. (1983) 'Origins of the underdevelopment of the Southern Sudan: British administrative neglect', Khartoum: University of Khartoum, Development Studies and Research Centre, Monograph Series no. 16.

Barnett, A. (1977) *The Gezira Scheme*, London: CASS.

Beshai, A. (1976) *Exports and Economic Development*, London: Ithaca.

Biggs, J. D. (1981) *Institutions and Decision Making in Agricultural Research*, London: Overseas Development Institute, Agricultural Administration Network, Discussion Paper no. 6, April 1981.

Bohannan, P. and Bohannan, L. (1968) *Tiv Economy*, London: Longman.

Boserup, E. (1965) *The Conditions of Agricultural Growth*, London: Allen & Unwin.

Cahill, K. M. (ed.) (1982) *Famine*, New York: Orsis.

CERES (1985) FAO In Action Choosing Priorities in Faculty Management, no. 104 (vol. 18, no. 2) March-April 1985, Rome.

Chambers, R. (1975) *Managing rural development: Ideas and Experience from East Africa*, Uppsala: The Scandinavian Institute of African Studies.

Clarke, J. (1986) *Resettlement and Rehabilitation: Ethiopia's Campaign Against Famine*, London: Harney & Jones.

Clay, J. W. and Holcomb, B. K. (1985) *Politics and the Ethiopian Famine 1984–85*, Cultural Survival:, Cambridge, Mass.

Cohen, J. (1978) *Land Tenure and Rural Development in Africa*, Cambridge, Mass.: Harvard Institute for International Development, Discussion Paper no. 44.

Currie, J. M. (1981) *The Economic Theory of Agricultural Land Tenure*, Cambridge: Cambridge University Press.

Curtis, D. (1988) 'Development administration in the context of world economics recession: some ideas on services provision in Southern Sudan', *Public Administration and Development* 8(1), pp. 2–13.

Dandekar, V. M. and Pethe, V.P. (1972) 'A survey of famine conditions in the affected regions of Maharashtra and Mysore', Poona, India: Gorhale Institute of Politics and Economics, Mimeograph Series no. 13.

De Waal, A. et al. (1986) *Report on Save the Children Fund Activities in Darfur*, Khartoum: Save the Children Fund.

Dodwell, H. H. (ed.) (1929) *The Cambridge History of India, vol. 5: British India 1497–1858*, Cambridge: Cambridge University Press.

Faisal Islamic Bank (1984) *Facts about Dura in the Faisal Islamic Bank during Season 1983/4*, Khartoum: FIB (in arabic).

Famine (1985) *Famine a Man Made Disaster* (a report for the Independent Commission on International Humanitarian Issues), London: Pan Books.

FAO Trade Yearbook, Rome: FAO.

FAO Production Yearbook, Rome: FAO.

FAO (1985) *Food and Agriculture Situation in African Countries Affected by Calamities in 1983–85*, Rome: FAO. Situation Report no. 7.

Feachem, R. et al. (1978) *Water Health and Development*, London: Trimed Books.

Feeney, D. (1984) *The Development of Prosperity in Land: A Comparative Study*, Paris: Economics Growth Centre, Centre Discussion Paper.

Foley, G. and Barnard, G. (1984) *Farm and Community Forestry*, London: Earthscan, Technical Report no. 3.

Francis, P. (1984) 'For the use and common benefits of all Nigerians: consequences of the 1978 land nationalisation', *Africa* 34 (3): 5–28.

Geertz, C. (1963) *Agricultural Involution: The Processes of Ecological Change in Indonesia*, Berkeley: University of California.

Government of Botswana (1981) *The management of Communal Grazing in Botswana*, Ramatlabame Range Management Centre, Botswana: Evaluation Unit, March.

Government of Gujarat, Revenue Department (1976) *The Gujarat Relief Manual*, vol. 1, Ghandinagar, India.

Government of Gujarat, Revenue Department (1982) *The Gujarat Relief Manual*, vol. 2, Ghandinagar, India.

Government of Sudan, Ministry of Agriculture, Food, and Natural Resources (1976) *Desert Encroachment Control and Rehabilitation Programme*, Khartoum.

Government of Sudan, Ministry of Agriculture, Food, and Natural Resources (1985) *Crop Production, 1985/6 Season: Preliminary Estimates*, Khartoum.

Government of Sudan, Ministry of Agriculture, Food, and Natural Resources, *Yearbook of Agricultural Statistics*, Khartoum.

Grandin, B. E. (1986) 'Land tenure, subdivision and residential change on a Masai group ranch', *Development Anthropology Network*, New York: Institute of Development Anthropology.

Gray, J. (1978) 'Mao Tse Tung's strategy in the collectivisation of Chinese agriculture', I. Oxaal, T. Barnett, and D. Booth (eds) *Beyond the Sociology of Development*, London: Routledge and Kegan Paul.

Grindle, M.S. (1980) 'Policy content and context in implementation', in M. S. Grindle (ed.) *Politics and Policy Implementation in the Third World*, Princeton: Princeton University Press.

Hales, J. (1979) *The Pastoral System of the Meidob*, Cambridge: Cambridge University Press.

Heaver, R. (1982) *Bureaucratic Politics and Incentive in the Management of Rural Development*, Washington, DC: World Bank Working Paper no. 537.

Hill, P. (1963) *The Migrant Cocoa Farmers of Southern Ghana: A Study in Rural capitalism*. Cambridge: Cambridge University Press.

Hobden, A. (1973) *Land Tenure and Rural Development in Africa*, Cambridge, Mass.: Harvard Institute for International Development, Discussion Paper no. 44.

Holt, P. M. (1950) *The Mahdist State*, Oxford: Oxford University Press.

Hunt, D. (1984) *Land and Inequality in Kenya*, Aldershot: Gower.

Ibrahim, F. (1984) *Ecological Imbalance in the Republic of Sudan – with Reference to Desertification in Darfur*, Bayreuth: Druckhaus Bayreuth Veragsgesellschaft.

International Monetary Fund (1985) *Balance of Payments Yearbook*, Washington, DC. IMF.

Jackson, A. J. K. (1976) 'The family entity and famine among the 19th century Akamba of Kenya: social responses to environmental stress', *Journal of Family History* 2 (winter): 193–216.

Kasfir, N. (1976) *The Shrinking Political Arena: Participation and Ethnicity in African Politics – with a Case Study of Uganda*, Berkeley: University of California Press.

Korten, D. C. (1984) 'Strategic organisation for people centered development', *Public Administration Review*, July/August.

Koshy, C. K. (1986) *Drought Management: The Gujarat Experience*, Ghandinagar, India: Office of the Commissioner of Relief (mimeo).

Leonard, D. K. (1985) 'African practice and the theory of user fees', *Agricultural Administration* 18: 137–57.

Loveday, A. (1914) *The History and Economics of Indian Famines*, London: G. Bell & Sons.

Mamdani, Md (1976) *Politics and Class Formation in Uganda*, London: Heinemann.

Mathur, K. and Bhattacharya, M. (1978) *Administrative Response to Emergency*, Delhi: Concept Publishing.

Mohamed, Y. et al. (1982) *Baseline Survey and Monitoring Programme for North Kordofan Rural Water Supply Project*, Khartoum: University of Khartoum, Institute of Environmental Studies.

National Records Office, London. Civ. Sec. 19/1/1–3, Famine Files.

Nyerere, J. (1984) 'Efficient farming for Tanzania', *Development and Cooperation* 6, pp. 12–14, Bonn: German Foundation for International Development.

O'Brien III, JJ. (1980) 'Agricultural labour and development in Sudan', University of Connecticut PhD thesis.

O'Dell, M. J. (1982) *Local Institutions and Management of Communal Resources: Lessons from Africa and Asia*, London: Overseas Development Institution, Agricultural Administration Unit. Pastoral Network Paper 14c.

Ohrwalder, J. (1895) *Ten Years' Captivity in the Mahdi's Camp, 1882–1892*, London.

Oxfam, UNICEF, and Darfur Regional Government (1985a) Reports on project activities (March–May), El Obeid, Sudan.

Oxfam, UNICEF, and Kordofan Regional Government (1985b) Reports on nutrition surveillance and drought-monitoring project (3 reports), El Obeid, Sudan.

Pacey, A. and Cullis, A. (1986) *Rainwater Harvesting: The Collection of Rainfall and Run-off in Rural Areas*, London: Intermediate Technology Development Group.

Peters, T. and Waterman, R. H. J. (1982) *In Search of Excellence: Lessons from America's Best Run Companies*, New York: Harper & Row.

Rangasami, A. (1985) 'Failure of exchange entitlements' theory of famine: a response', *Economic and Political Weekly* XX, 41, October 12, 19.

Richards, P. (1985) *Indigenous Agriculture Revolution: Ecology and Food Production in West Africa*, London: Hutchinson.

Sen, A. K. (1981) *Poverty and Famines: An Essay on Entitlement and Deprivation*, Oxford: Oxford University Press.

Sen, A. K. (1983) 'Development: which way now?' *Economic Journal* (December): 745–62.

Shepherd, A., Norris, M., and Watson, J. (1987) *Water Planning in Arid Sudan*, London: Ithaca.

Shepherd, G. (1985) *Social Forestry in 1985: Lessons Learnt and Topics to be Addressed*, London: Overseas Development Institute Social Forestry, ODI Network Paper Ia.

Simpson, I. and Simpson, M. (1978) *Alternative Strategies for Agricultural Development in the Central Rainlands of the Sudan*, Leeds: University of Leeds, Rural Development Study no. 3.

Sudanow (1983) *Sudan Yearbook 1983*, Khartoum: Government of Sudan.

Suliman, el A. (1984) 'The contribution of Faisal Islamic Bank in the development of Sudan', Khartoum University, MSc thesis.

Swift, J. (1974) 'Drought and a Sahelian nomad economy', in D. Dalby and H. Church (eds), *Drought in Africa*, London: University of London, School of Oriental and African Studies, Centre for African Studies.

Tickner, V. (1985) 'Military attacks, drought and hunger in Mozambique', *Review of African Political Economy* 33 (August): 89–91.

Timberlake, L. (1985) *Africa in Crisis*, London: Earthscan.

Van Apeldoorn, G. J. (1981) *Perspectives on Drought and Famine in Nigeria*, London: Allen & Unwin.

Weinbaum, M. F. (1982) *Food, Development and Politics in the Middle East*, London: Croom Helm.

World Bank (1981) *Accelerated Development in Sub-Saharan Africa* (Berg Report), Washington, DC: World Bank.

World Bank (1984) *Towards Sustained Development in Sub-Saharan Africa: A Joint Programme of Action*, Washington, DC: World Bank.

World Bank (1986a) *Poverty and Hunger: Issues and Options for Food Security in Developing Countries*, Washington, DC: World Bank.

World Bank (1986b) *Financing Adjustment with Growth in Sub-Saharan Africa*, Washington, DC: World Bank.

Index

Famine Code, 6, 111, 122–3, 144, 145,
150, 161, 165, 169
Famine Commission, 121–2
famine prevention, 3–4; aid agencies
(role), 226–31; financing (Africa), 25–7;
government role, 215–31; NGOs (role),
215, 231–3; policies, 6–8, 11–13;
research and development, 234–6
Famine Regulations, 36, 39–41, 66–71
farm forestry, 186
Farming Systems Research, 188, 189–90
fattening schemes, 175
Feachem, R., 198, 204
Feeney, D., 200
fertilizers, 120, 158, 160, 165, 179, 183,
185–7
feudal society, 77–82, 106
financing: famine prevention, 25–7; self-,
23
Firebrace, J., 108
Five-Year Plans, 123, 159
Flood, G., 109
floods, 111, 131–6
fodder policy, 128
Foley, G., 208, 209
food: availability decline, 60, 61–2;
consumption, 13, 14, 18–20; crisis
management, 131–6; economy (Africa/
India), 167–8; entitlement programmes,
141–55; production, 13–14, 16–20,
167–8; riots, 183; stamps, 142, 143;
subsidies, *see* subsidies; supply, 14, 15,
145–50; system, 132–3, 134; wage,
148–50
Food-for-Work: advantages/problems,
148–50; Bangladesh, 133–6, 151;
Botswana, 119, 222; implications (long-
run), 153–4; India, 159, 165; South
Asia, 141, 144, 145, 153–4
Food and Agriculture Organization, 13,
14, 15–17, 91–2, 171, 177
food aid: agencies, *see* aid agencies;
Ethiopia, 89–96, 106–7; food-imports
trap, 167–8; Sudan, 28, 53–9, 64–5;
volume of, 17–18, 23
Food Aid National Administration, 53, 56
food distribution, 144: Africa, 167–8;
Botswana, 115–16, 117; Gujarat, 130–1,

134; India, 158–9, 163–4, 167–8;
Sudan, 50–1, 52–3; *see also* transport
capacity
food security, 6–8: emergency measures
(India), 157–70; emergency measures
(South Asia), 141–55; rural
development and, 152–3
foreign exchange, 7, 52, 113, 135, 182,
185, 191, 228
Forest Conservation Act (1980), 160
forestry, 194, 208–9; deforestation, 33, 60,
65, 102, 163
Francis, P., 204
free market, 42, 84, 104, 200, 226
free press, 161, 165

Gash Delta Scheme, 52–3
Gavan, J.D., 155
Gazette of the Government of India, 122
Geertz, C., 180
George, P.S., 155
Gezira Scheme, 192
Ghassemi, H., 156
Gill, P., 109
Gizaw, Berhane, 89
GNP (famine-indicator), 20–1, 23–4
Gonder liberation movement, 75, 77
government (role): Africa/India
(comparison), 166–70; Bangladesh,
131–6; Botswana, 113, 114–16;
entitlement intervention, 141–55;
Ethiopia, 105–7; famine prevention
role, 215–31; Gujarat, 121–31; India,
157–65; India/Africa (comparison),
166–70; legal reforms, 199–200;
livestock policy, 173–6; research and
development, 234–6; Sudan, 52–5, 58,
63–4
Government of Gujarat, 12, 122, 123, 124,
126, 129
Government of Sudan, 34, 62
Grain Deficit Study Committee, 81
Grandin, B.E., 210
grass (control/use), 210–11; *see also*
grazing schemes
gratuitous relief, 129
Graves, J., 155, 156
Gray, J., 200